MW00843954

INCREDIBLE
FACTS

Publications International, Ltd.

Images from Shutterstock.com

Copyright © 2023 Publications International, Ltd. All rights reserved. This book may not be reproduced or quoted in whole or in part by any means whatsoever without written permission from:

Louis Weber, CEO
Publications International, Ltd.
8140 Lehigh Avenue
Morton Grove, IL 60053

Permission is never granted for commercial purposes.

ISBN: 978-1-63938-411-2

Manufactured in China.

8 7 6 5 4 3 2 1

Let's get social!

 @Publications_International

 @PublicationsInternational

www.pilbooks.com

CONTENTS

Chapter 1

PLANET EARTH

33 Amazing Earth Facts

1. Earth weighs approximately 5,940,000,000,000,000,000,000 metric tons.

2. The surface of Earth is approximately 70.9 percent water and 29.1 percent land.

3. Earth is the only planet not named after a god.

4. The deepest place on Earth is Challenger Deep in the Mariana Trench, with a depth of 35,840 feet below sea level.

5. Earth is not a perfect sphere, more of a squished ball.

6. Thanks to Earth's quirky shape, Mount Chimborazo, which rises only 20,702 feet from sea level, is about a mile and a half closer to the moon than Everest's peak (29,035 feet from sea level).

7. If you stood at the equator, you would weigh slightly less than if you stood at the North or South Pole.

8. The distance around the equator is 24,901 miles.

9. Earth is fatter in the middle near the equator where gravity pushes to create a bulge.

10. Planet Earth is in constant motion. This movement causes mountains to rise, earthquakes to rumble, and volcanoes to spew out hot rock.

11. Made of solid iron and nickel, Earth's inner core is as hot as the surface of the sun—about 10,000 °F.

12. It doesn't melt, though, because it's compressed by the weight of the planet all around it.

13. Surrounding the inner core is the outer core. This, too, is mostly iron and nickel, and very hot.

14. The outer core's rock is liquid because it's not under as much pressure as the inner core.

15. Around the outer core is the mantle. It is the thickest part of Earth, about 1,800 miles deep.

16. The mantle is solid rock, but so hot that it flows like a thick, gummy liquid.

17. Earth's thin outermost layer is the crust, which covers the mantle.

18. Veryovkina Cave in the country of Georgia is the world's deepest cave, plunging 7,257 feet.

19. Earth is the only planet in our solar system to have water in its three states of matter: liquid, solid, and gas.

20. The Caspian Sea, situated where southeastern Europe meets Asia, is the world's largest lake with a surface area of 143,243 square miles.

21. Stretching for more than 270 miles, the Grand Canyon in Arizona is the world's largest canyon.

22. The Himalayas are the tallest and youngest mountains on Earth.

23. Today's continents were split apart from a super continent named Pangea.

24. Bouvet Island in the South Atlantic is Earth's most remote island.

25. It lies almost a thousand miles from the nearest land (Queen Maude Land, Antarctica).

26. Earth is thought to be about 4.54 billion years old.

27. Compared to geologic time, humans have lived on Earth for the blink of an eye.

28. Chicxulub Crater in the Yucatán Peninsula is the largest confirmed impact crater on Earth.

29. The asteroid that formed this 105-mile-diameter crater is widely credited with killing off the dinosaurs.

30. Earth is moving through the solar system at around 67,000 miles per hour.

31. If the sun suddenly stopped emitting light and warmth, Earth would get dark in about eight minutes—the length of time it takes for light to reach us once it escapes the sun—and would gradually become colder.

32. There are 195 countries in the world.

33. Russia is the world's largest country by area.

12 Blazing Hot Facts

1. The world's hottest temperature of 134 °F was recorded on July 10, 1913, in Furnace Creek, Death Valley, California.

2. The Sahara is the world's largest hot desert.

3. Despite the popular saying, it's never hot enough to fry an egg on the sidewalk.

4. The pavement would need to hit at least 158 °F for the egg to cook, and even blacktop only reaches 145 °F.

5. Hot water should boil at 212 °F (100 °C), but the actual boiling point varies depending on atmospheric pressure, which changes according to elevation.

6. A typical lightning bolt is hotter than the surface of the sun.

7. The Chicago heat wave of 1995 killed 739 people, mostly poor elderly residents who couldn't afford air conditioning.

8. The Red Sea is the world's hottest. It's 87 °F at its warmest spot.

9. In Death Valley, California, the average monthly temperature for July is 101 °F.

10. Phoenix, Arizona, recorded 128 days with a temperature of 100 °F or higher in 2018.

11. The city had 143 days with triple digit temps in 1989.

12. Massive heat waves across North America were persistent in the 1930s during the Dust Bowl, killing at least 5,000 people in the U.S. alone.

32 Cold Hard Facts

1. Water is supposed to freeze at 32 degrees Fahrenheit (0 degrees Celsius), but scientists have found liquid water as cold as −40 °F in clouds and even cooled water down to −42 °F in the lab.

2. The world's lowest temperature of −128.5 °F was recorded at Vostok Station, Antarctica, on July 21, 1983.

3. Icebergs are formed from glaciers on land and drift out to sea. They are mostly made from freshwater, not saltwater.

4. Snowflakes are collections of ice crystals.

5. The way that ice crystals join together gives every snowflake a unique design.

6. One snowflake can contain as many as 100 ice crystals.

7. What shape a snowflake takes depends on the temperature and the amount of moisture in the cloud.

8. In 1988, two identical snowflakes collected from a Wisconsin storm were confirmed to be twins at an atmospheric research center in Colorado.

9. The Rainier Paradise Ranger Station in Washington State gets an average 676 inches of snow annually.

10. Earth's two ice sheets cover most of Greenland and Antarctica. They make up more than 99 percent of the world's glacial ice.

11. Most North Atlantic icebergs calve (detach) from the Greenland ice sheets.

12. "Growlers" are mini-icebergs and "bergy bits" are big ice chunks.

13. Ninety percent of Greenland is covered by an ice cap and smaller glaciers.

14. The coldest temperature ever recorded in North America was −87.0 °F on January 9, 1954, at the North Ice research station on Greenland.

15. Harbin, China, often called the "ice city," hosts the Harbin International Snow and Ice Festival each year.

16. Temperatures as low as −44 °F have been recorded in Harbin.

17. The village of Oymyakon, Russia, is the coldest permanently occupied human settlement in the world (population 500).

18. During the winter, temperatures average −58 °F and 21 hours per day are spent in darkness.

19. The all-time coldest temperature in Oymyakon, −98 °F, was recorded in 2013.

20. International Falls, Minnesota, took another town to court over the title of coldest place in the continental United States.

21. Temperatures as low as −40 °F have been recorded there.

22. The title of coldest place in the continental United States actually belongs to Stanley, Idaho, with a record cold temperature of −52.6 °F.

23. Not all blizzards have falling snow.

24. Ground blizzards occur when no snow is falling.

25. Ulaanbaatar, Mongolia, is the coldest capital city.

26. The average annual temperature is 29.6 °F in Ulaanbaatar, but temperatures range between 5 °F and −40 °F in winter months.

27. While it seems counterintuitive, Earth is actually closer to the sun during winter in the Northern Hemisphere.

28. When you hear thunder during a snowstorm, it's called thundersnow.

29. Lightning is harder to see in the winter and the snow sometimes dampens the thunderous sound.

30. Snow falls one to six feet per second.

31. More than 22 million tons of salt are used to de-ice U.S. roads each winter.

32. The average monthly temperature for January in Verkhoyansk, Russia, is −53 °F.

36 Facts About North America

1. North America is the world's third largest landmass with an area of 9,449,078 square miles.

2. Only about eight percent of the world's population live in North America.

3. The United States is the world's third most populous country with an estimated 334,233,854 people.

4. North America has no landlocked countries.

5. Central America refers to the seven countries of North America between Mexico and South America: Belize, Costa Rica, El Salvador, Guatemala, Honduras, Nicaragua, and Panama.

6. Honduras is the only Central American country without an active volcano.

7. Mexico City, Mexico, is the highest North American capital with an elevation of 7,349 feet.

8. North America was first populated about 10,000 years ago when people moved across the Bering Sea between Siberia and Alaska.

9. Bermuda is the most densely populated country in North America, with 3,200 people per square mile.

10. Groups of mountain quail (*Oreortyx pictus*), native to western North America, often walk up to 5,000 feet in winter in the longest bird migration on foot.

11. Guatemala's national bird is the quetzal.

12. Canada is the world's second largest country with a total area of 3,855,102 square miles.

13. The world's biggest island, Greenland, is part of North America, although it belongs to the European country of Denmark.

14. Canada is the world's largest country that borders only one other country (United States).

15. Sinkholes called *cenotes* on Mexico's Yucatán Peninsula were sacred to the ancient Maya.

16. Nearly four million people of Maya descent still live in southern Mexico and Central America.

17. North America's longest mountain chain is the Rocky Mountains, which stretch 3,000 miles across the United States and Canada.

18. Grenada, nicknamed the Spice Island, is one of the world's leading sources of nutmeg and other spices.

19. The Panama Canal divides North and South America.

20. The world's longest shared border is the 3,987-mile boundary between the United States and Canada. That doesn't include Canada's 1,539-mile border with Alaska.

21. Approximately 90 percent of Canadians live within 100 miles of the U.S. border.

22. Alaska's Denali (Mount McKinley) is North America's highest point at 20,308 feet.

23. Grant's caribou (*Rangifer tarandus granti*) of Alaska and the Yukon Territory of North America travel up to 2,982 miles per year.

24. Lake Superior is North America's largest lake and one of the largest freshwater lakes on the planet.

25. Canada has up to three million lakes—more than all other countries combined.

26. Bracken Cave outside of San Antonio, Texas, is the world's largest bat cave. An estimated 20 million Mexican free-tailed bats roost in the cave from March to October.

27. The Yosemite Falls are North America's highest waterfalls, dropping 2,425 feet.

28. Hawaii's Mauna Kea is the world's tallest mountain as measured from base to summit.

29. Its base begins 19,685 feet below sea level; total height estimates range from 32,696 to 33,474 feet.

30. The Canadian Arctic Archipelago consists of 36,563 islands.

31. Canada has the longest coastline of any country in the world by far.

32. Off the coast of Belize in the Caribbean Sea is the Great Blue Hole, a submarine sinkhole measuring more than 1,000 feet across and 400 feet deep.

33. Haiti and the Dominican Republic share the island of Hispaniola.

34. Cuba is the largest Caribbean island with a total area of 42,803 square miles.

35. The Mississippi River runs 2,202 miles from Minnesota down to the Gulf of Mexico.

36. An earthquake on December 16, 1811, caused parts of the Mississippi River to flow backward.

33 Bridge Facts

1. With a total length of 102 miles, the Danyang-Kunshan Grand Bridge, on the Jinghu High-Speed Railway in China, is the world's longest bridge.

2. There are no bridges that span the entire Amazon River.

3. There are more than 600,000 bridges in the United States.

4. China's Zhaozhou (or Anji) Bridge is the world's oldest stone segmental arch bridge.

5. Built around A.D. 600, it is still standing today.

6. The Rockville Bridge in Harrisburg, Pennsylvania, is the longest stone arch bridge in the world.

7. It was built in 1902 and measures 3,820 feet.

8. Fifteen major bridges span the waters of the Allegheny and Monongahela rivers as they flow together to become the Ohio River.

9. The Sydney Harbour Bridge in Australia can rise or fall up to seven inches depending on the temperature due to the steel expanding or contracting.

10. One of the longest steel-arch bridges in the world, it features six million rivets.

11. California's Golden Gate Bridge spans the narrow body of water between San Francisco Bay and the Pacific Ocean.

12. Despite its name, the famous suspension bridge is painted a reddish orange, not gold.

13. The bright color is called "international orange."

14. The main span of the Golden Gate Bridge is 4,200 feet long.

15. It was the longest bridge in the world from 1937 to 1964.

16. During the construction, 11 workers fell from the bridge and died, though a safety net saved the lives of 19 others.

17. The Charles Bridge, which spans the Vltava River, in Prague, Czech Republic, dates from 1357.

18. The stone bridge is lined with statues and has towers with gates at each end.

19. Turkey has the world's longest cable suspension bridge. The main span of the 1915 Çanakkale Bridge is 6,637 feet, or 1.25 miles.

20. The total length of the Çanakkale Bridge is 11,690 feet, or 2.21 miles. It spans the Dardanelles strait.

21. Before Turkey's Çanakkale Bridge opened in 2022, Japan's Akashi-Kaikyo bridge held the record for world's longest cable suspension bridge.

22. The main span of the Akashi-Kaikyo bridge is 6,532 feet, or 1.24 miles, long.

23. The nearly five-mile-long Mackinac Bridge links Michigan's Upper and Lower Peninsulas.

24. The Brooklyn Bridge, designed by architect John Augustus Roebling, towers over New York City's East River.

25. It opened on May 24, 1883, linking what would become the boroughs of Brooklyn and Manhattan in New York City.

26. In its day, the Brooklyn Bridge was the longest suspension bridge in the world, with the length of the main span measuring 1,595 feet.

27. The Øresund Bridge, connecting Sweden with Denmark, is part overwater and part underwater tunnel.

28. The New River Gorge Bridge is one of the most photo-graphed places in West Virginia.

29. The bridge in the Appalachian Mountains is one of the highest bridges open to vehicles in the world.

30. The New River Gorge Bridge cut the travel time across the gorge from an arduous 45 minutes on mountain roads to an easy minute across the bridge.

31. The world's longest steel truss-arch bridge is the Chongqing-Chaotianmen Bridge over the Yangtze River in China.

32. The main span is 1,811 feet long.

33. Oklahoma City honored its state bird, the scissor-tailed flycatcher, with the design of the Skydance Bridge.

11 Facts About Giant's Causeway

1. Giant's Causeway is Northern Ireland's only UNESCO World Heritage Site.

2. Over 40,000 interlocking basalt columns make up this geological formation.

3. The columns of Giant's Causeway typically have five to seven irregular sides.

4. Most are hexagonal in shape.

5. The strange landscape was the result of an ancient vol-canic fissure eruption during the Paleogene Period.

6. The tallest columns in Giant's Causeway are approxi-mately 39 feet high.

7. The site also notes more mythological origins: Long ago, the Irish giant Fionn mac Cumhaill tore up the Antrim coastline during a battle with Scottish giant Benandonner. As he hurled the rocks into the water, he paved a route over the sea to Scotland.

8. Other popular examples of columnar jointing include Fingal's Cave in Scotland, Columbia River flood basalts in Oregon, Devils Postpile in California, and Devil's Tower in Wyoming.

9. German composer Felix Mendelssohn based his 1830 "Hebrides Overture" on the sound of waves flowing into Fingal's Cave.

10. Led Zeppelin's album cover for *Houses of the Holy* features Giant's Causeway.

11. Nicknames for several of the formations in this peculiar landscape include "the organ pipes," "the honeycomb," and "the chimneys."

43 Wild Weather Facts

1. All weather occurs within eight miles of Earth's surface.

2. All snowflakes have six sides or points.

3. The world's greatest 24-hour rainfall record was set in January 1966 on the French island of Réunion in the Indian Ocean.

4. A record 71.8 inches of rain fell.

5. Tropical storms rotate clockwise south of the equator and counterclockwise in the Northern Hemisphere.

6. A hurricane is given a rating on the Saffir-Simpson scale based on its sustained wind speed.

7. The Enhanced Fujita Scale (EF-Scale) is a system for classifying tornado intensity.

8. Waterspouts are tornadolike rotating columns of air over water.

9. A tropical storm must have winds of 74 miles per hour or greater to be classified a hurricane.

10. Tornadoes have been reported on all continents except Antarctica.

11. Contrary to popular belief, tornadoes can and do hit downtown areas of major cities.

12. The city center of St. Louis, Missouri, has been flattened four times since 1871.

13. All thunderstorms produce lightning, which is what makes the sound of thunder.

14. The world's heaviest hailstones weighed up to two pounds each and killed 92 people in Gopalganj district, Bangladesh, on April 14, 1986.

15. The United States has more tornadoes than any other country.

16. The Great Blizzard of 1888 blanketed some areas of the Atlantic coast with as much as 50 inches of snow.

17. Waterspouts can make sea creatures rain down from the sky.

18. The eye of a hurricane is calm, but the eye wall that surrounds it has the strongest, most violent winds.

19. Snowflakes formed in clouds take about an hour to reach the ground.

20. Hurricanes, tropical cyclones, and typhoons all describe the same type of storm—what they're called depends on where they form.

21. When lightning strikes, the heated air creates a sonic shock wave, which is the sound of thunder.

22. We see lightning before we hear thunder because light travels much faster than sound.

23. The top speed of a falling raindrop is 18 miles per hour.

24. Crickets make good thermometers—to get a rough estimate of the temperature, count the number of times a cricket chirps in 15 seconds and add 40.

25. Tornadoes usually spin counterclockwise in the Northern Hemisphere and clockwise in the Southern Hemisphere.

26. The winds of a tornado are the strongest on Earth. They may reach a speed of 300 miles per hour.

27. Most hurricanes stretch about 300 miles across.

28. The Tri-State Tornado was the deadliest tornado in U.S. history.

29. It traveled through Missouri, Illinois, and Indiana on March 18, 1925, killing 695 people.

30. Dirty snow melts faster than clean snow.

31. The Great Blizzard of 1888 convinced many people that passenger trains should run underground and helped to inspire cities like Boston and New York to build the first subway systems.

32. The fastest wind on Earth blew through the suburbs of Oklahoma City, Oklahoma, on May 3, 1999.

33. The 301-mile-per-hour gusts were recorded during an EF5 tornado that destroyed hundreds of homes.

34. The most active Atlantic hurricane season on record was in 2020.

35. National Hurricane Center ran out of names and had to use Greek letters for the last several storms.

36. The Tri-State Tornado lasted 3.5 hours and traveled 219 miles—setting records for both duration and distance traveled.

37. Low barometric pressure generally indicates stormy weather, and high pressure signals calm, sunny skies.

38. The lowest barometric pressure ever recorded was 25.69 inches (870 mb) during Typhoon Tip in 1979.

39. The highest barometric pressure, 32.01 inches (1,083.8 mb), was measured in 1968 on a cold New Year's Eve in northern Siberia.

40. A tropical storm officially becomes a hurricane, cyclone, or typhoon when winds reach at least 74 miles per hour.

41. The deadliest storm on record was the Bhola Cyclone, which hit East Pakistan (now Bangladesh) on November 13, 1970, killing at least 500,000 people.

42. The Bhola Cyclone's winds were in excess of 120 miles per hour when it finally hit land.

43. It generated an astonishing storm surge of 12 to 20 feet, which flooded densely populated coastal areas.

50 South America Facts

1. South America is home to the Andes, the world's longest continental mountain range.

2. At 4,700 miles in length, the Andes span seven countries.

3. South America has 12 independent countries.

4. Two territories in South America are still administered by European powers: French Guiana (France) and the Falkland Islands (United Kingdom).

5. Easter Island, a Chilean territory, is some 2,300 miles from the coast of South America.

6. Brazil is the largest country in South America and in the Southern Hemisphere with a total area of 3,287,957 square miles.

7. Colombia has coastlines on both the Pacific and Atlantic Oceans.

8. The Galápagos, a volcanic chain of islands belonging to Ecuador, is home to the world's only marine iguanas.

9. Suriname is the smallest country in South America with a total area of 63,251 square miles.

10. Both North and South America are named after Italian explorer Amerigo Vespucci.

11. La Paz, Bolivia, is the world's highest administrative capital at 11,942 feet above sea level.

12. Ecuador is the most densely populated country in South America with 135 people per square mile.

13. The Pantanal region of South America is the world's largest freshwater wetland.

14. It is almost ten times the size of the Florida Everglades.

15. The potato originates in South America.

16. The Amazon River runs through South America and is surrounded by the world's largest rain forest.

17. The world's highest gas pipeline tops out at over 16,077 feet in the Peruvian Andes.

18. The world's largest river basin is that drained by the Amazon, which covers about 2,720,000 square miles.

19. The Andean condor has a wingspan up to 10 feet 10 inches—one of the largest wingspans of any bird.

20. The Uyuni Salt Flats cover more than 4,000 square miles of Bolivia at an altitude of 12,500 feet.

21. Because of brine just below the surface, any crack in the salt soon repairs itself.

22. Guyana has up to 300 species of catfish.

23. The Incas were the largest group of indigenous people in South America.

24. The Incan Empire lasted from 1438 until 1533.

25. Brazil's men's soccer team boasts the most World Cup wins with five titles.

26. Cerro Aconcagua, an Andean peak in Argentina, is South America's highest point at 22,834 feet above sea level.

27. Brazil shares a border with every other South American country except Chile and Ecuador.

28. Spanning the border between Peru and Bolivia, Lake Titicaca is South America's largest lake at 3,200 square miles.

29. Uruguayan cowboys are called "gauchos."

30. Argentina's president works in a pink building called the Casa Rosada ("Pink House").

31. The Itaipu Dam between Brazil and Paraguay is the world's second-largest hydroelectric facility.

32. It supplies three-quarters of the electricity used in Paraguay.

33. Venezuela's Angel Falls is the world's highest waterfall.

34. The main free-falling segment is almost 20 times higher than Niagara Falls.

35. Argentina and Chile share South America's largest island: Tierra del Fuego.

36. The southernmost city in the world is Ushuaia, located on the Argentinian part of the Tierra del Fuego.

37. Brazil is the most populous South American country with an estimated 214.3 million people.

38. Overlooking Rio de Janeiro, Brazil, stands a colossal statue of Jesus Christ called *Christ the Redeemer*, the world's largest Art Deco-style sculpture.

39. Suriname is the least populous South American country with an estimated 612,985 people.

40. Chile is a narrow strip on the southwestern edge of South America.

41. It offers 3,998 miles of South Pacific coastline.

42. Built by an Inca ruler between 1460 and 1470, Machu Picchu reveals the Inca's skills as stone masons.

43. The Amazon River carries more water than any other river on the planet.

44. The Atacama Desert in Chile is the world's driest non-polar desert.

45. Parts of the central desert area regularly go without rain for years at a time.

46. Approximately half of South America's population lives in Brazil.

47. The ancient kingdom of Tiwanaku was a major Indian civilization in the Andes Mountains of South America.

48. Aztec Emperor Montezuma is said to have drunk 50 goblets of chocolate, flavored with chili peppers, every day.

49. The world's widest road is Brazil's Monumental Axis.

50. It could hold 160 cars side-by-side.

22 Earth-quaking Facts

1. Earthquakes occur when a fault (where Earth's tectonic plates meet) slips, releasing energy in waves that move through the ground.

2. Earth has 16 tectonic plates.

3. Earthquakes occur almost continuously. Fortunately, most are undetectable by humans.

4. An estimated 500,000 earthquakes are detected in the world each year.

5. Scientists measure earthquake magnitude based upon the amount of energy released by the rock movements.

6. An earthquake with a Richter magnitude of 2 is about the smallest earthquake that can be felt by humans without the aid of instruments.

7. The highest magnitude earthquake ever recorded, a 9.5, was centered off the coast in Chile in 1960.

8. Seismographs recorded seismic waves that traveled all around the Earth for days after the Chilean earthquake in 1960.

9. More earthquakes occur in Alaska than in any other U.S. state.

10. The San Andreas Fault, an 800-mile fracture in Earth's crust that stretches along the California coast, moves at the same rate your fingernails grow—about 2 inches per year.

11. If that rate keeps up, scientists project that Los Angeles and San Francisco will be adjacent to one another in approximately 15 million years.

12. Shifting tectonic plates will someday cause the Horn of Africa to break apart from the rest of the continent.

13. Moonquakes are earthquakes on the moon.

14. Antarctica has icequakes, which are little earthquakes within the ice sheet.

15. Indonesia has more earthquakes than any other country.

16. The Loma Prieta earthquake, which struck the San Francisco area in 1989, delayed the World Series between the San Francisco Giants and the Oakland Athletics by 10 days.

17. The most powerful tremor in U.S. history—lasting three minutes and measuring 9.2 on the Richter scale—struck Prince William Sound in Alaska on March 28, 1964.

18. The December 2004 9.1-magnitude earthquake that struck off the coast of the Indonesian island of Sumatra, and the tsunami that followed, killed at least 230,000 people.

19. Scientists say the tremor was so strong that the December 26, 2004, Indonesian quake wobbled Earth's rotation on its axis by almost an inch.

20. A seiche is the sloshing of water that happens during and after an earthquake. The swimming pool at the University of Arizona–Tucson lost water from sloshing caused by the 1985 Michoacán, Mexico, earthquake 1,240 miles away.

21. Seismology is the study of earthquakes.

22. Alaska experiences about 60 earthquakes per day.

48 Amazing Architecture Facts

1. Abraj al-Bait, also called Makkah Royal Clock Tower, is the world's tallest building with a clock face.

2. The multitowered skyscraper complex is adjacent to the Great Mosque in Mecca, Saudi Arabia.

3. The central clock tower, including its spire, rises to a height of 1,972 feet.

4. Taliesin is regarded as a prime example of Frank Lloyd Wright's organic architecture. The house and 600-acre estate is in Spring Green, Wisconsin.

5. The Sydney Opera House in Australia is a complex of theaters and halls linked beneath its famous white roof, the design of which suggests billowing sails.

6. The largest "sail" is as tall as a 20-story building.

7. Originally built for the 1962 World's Fair, Seattle's 605-foot Space Needle was the tallest building west of the Mississippi when it was completed.

8. The futuristic blueprints for the Space Needle evolved from artist Edward E. Carlson's visionary doodle on a placemat.

9. The resulting design looks like a flying saucer balanced on three giant supports.

10. The Space Needle was built to withstand winds of up to 200 miles per hour.

11. The Krzywy Domek (Crooked House) in Sopot, Poland, isn't a home, but part of the Rezydent shopping center.

12. Its wavy architecture looks like a fun house mirror reflection.

13. The Crooked House's bent lines and distorted walls and doors were inspired by the fairytale illustrations of Jan Marcin Szancer and Per Dahlberg.

14. The Empire State Building in New York City was the world's tallest building when it opened in 1931, soaring 1,454 feet to the top of its lightning rod.

15. More than 3,000 workers took less than 14 months to build the Empire State Building, the framework erected at a pace of 4.5 stories per week.

16. The pyramids of Giza, Egypt, are the only wonders of the ancient world that still stand.

17. They were built between about 2650 and 2500 B.C.

18. The Great Pyramid is the largest pyramid ever built.

19. It is 482 feet tall and contains some 2.3 million stone blocks, each weighing an average 2.7 tons.

20. At 1,483 feet, the Petronas Twin Towers in Kuala Lumpur, Malaysia, became the first skyscrapers to top any U.S. skyscrapers when they were completed in 1998.

21. Built in the 1740s, Drayton Hall in South Carolina is America's oldest unrestored plantation house that is open to the public.

22. Drayton Hall is considered one of the finest examples of Georgian Palladian architecture in the nation.

23. There are 10,000 panes of glass in California's Crystal Cathedral.

24. At 2,080 feet, Japan's Tokyo Skytree is the world's tallest freestanding communications tower.

25. The Colosseum in Rome, Italy, once seated 50,000 people.

26. Mazes beneath the main floor kept wild animals contained in preparation for events.

27. Trapdoors allowed gladiators to make surprise appearances and the dead bodies to be hauled away.

28. At 630 feet tall, the Gateway Arch in St. Louis, Missouri, is the nation's tallest monument. The distance between its two legs is equal to its height.

29. The towering steel structure was designed by Finnish-American architect Eero Saarinen in 1947 and completed in 1965.

30. The top of the Gateway Arch is accessible by two trams—one in each leg—that are made up of eight cylindrical, five-seat compartments.

31. The 110-story Willis Tower (formerly Sears Tower) in Chicago was the tallest building in the world from 1974 until 1998.

32. Willis Tower visitors can take a 45-second elevator ride to the 103rd floor, 1,353 feet up, and stand on the Ledge, a series of windows that extend from the building.

33. The Ledge is made of three layers of half-inch thick glass laminated into one seamless unit. It is built to withstand four tons of pressure and can hold 10,000 pounds.

34. In the ancient Mayan city of Chichén Itzá, Mexico, is the Temple of Kukulcán ("El Castillo").

35. Each side of the pyramid has a staircase made of 91 steps that leads to the top.

36. The 1,776-foot One World Trade Center in New York City is the tallest building in the Western Hemisphere.

37. The Shanghai Tower is the world's second-tallest building at 2,073 feet, after the 2,717-foot Burj Khalifa.

38. The Shanghai Tower's elevator travels 3,543 feet per minute.

39. Completed in 1439, the Strasbourg Cathedral in France was the world's tallest building until 1874.

40. The first skyscraper was pioneered in Chicago with the 138-foot-tall Home Insurance Building in 1885.

41. China has more skyscrapers than any other country.

42. India's Taj Mahal was commissioned by Shah Jahan to honor his wife Mumtaz, who died in childbirth.

43. It took 20,000 workers and 1,000 elephants 22 years to build the Taj Mahal. The structure and its grounds cover 42 acres.

44. Barcelona, Spain's Casa Milà, known as La Pedrera (or, "The Quarry"), expresses Antoni Gaudí's personal architectural style.

45. The façade of Casa Milà was built without flat surfaces, straight lines, or symmetry.

46. The roofline undulates, punctuated by chimneys that look squeezed from a pastry tube.

47. The terms "log house" and "log cabin" actually denoted two distinct types of dwellings. Log cabin timbers were left round while log houses were made from notched, square-hewn logs.

48. Cupboards originated with pioneers and were truly "cup boards"—a shelf built from a single board to hold cups and dishes. It was only later that they acquired sides and fronts.

18 Mariana Trench Facts

1. The Mariana Trench in the Pacific Ocean is the deepest location on planet Earth.

2. It is named after the nearby Mariana Islands, which themselves were named after the Spanish Queen Mariana of Austria.

3. Guam, a territory of the U.S., is less than 200 nautical miles away from the trench, giving the U.S. jurisdiction over it.

4. The trench was designated a U.S. national monument in 2009.

5. The Mariana Trench marks the tectonic boundary between two plates.

6. Because the Pacific plate is composed of older crustal material, it is colder and denser than the younger Mariana plate. As a result, it subducts underneath.

7. Mariana Trench's deepest point, called Challenger Deep, measures 36,201 feet deep. This is nearly 7 miles below sea level.

8. If you placed Mount Everest at the bottom of Challenger Deep, its peak would still be 7,000 feet underwater.

9. Crescent shaped, the trench is over 1,550 miles long and 43 miles wide.

10. Pressure increases with depth. At the bottom of the trench, the water column exerts a pressure of more than a thousand times the standard atmospheric pressure at sea level.

11. The depth and perpetual darkness of the Mariana Trench gives it a temperature just a few degrees above freezing: about 34 to 39 °F.

12. The first global oceanographic cruise, the 1875 H.M.S. *Challenger* expedition, used weighted sounding rope to measure the trench.

13. Later in 1951, H.M.S. *Challenger II* surveyed the trench again using echo sounding. This was an easier and more accurate way to measure depth.

14. 1960 was the first and only time humans have descended into the Challenger Deep. Jacques Piccard and Navy Lt. Don Walsh boarded the U.S. Navy bathyscaphe called the *Trieste*.

15. While their descent into the depths took five hours, only 20 minutes were spent exploring the bottom.

16. Due to the silt stirred up by their submersible, photography was impossible.

17. In 2012, Canadian filmmaker James Cameron piloted the *Deepsea Challenger* to a depth of 35,756 feet. This set a world record for a solo descent.

18. Data suggests that microbial life forms occupy the bottom of the Mariana Trench, and scientists believe there are many new species awaiting discovery.

54 Plant Facts

1. There are about 400,000 known plant species.

2. The giant corpse flower is named for its smell.

3. It secretes cadaverene and putrescine, odor compounds responsible for its rotting flesh smell.

4. The first type of aspirin came from the tree bark of a willow tree.

5. Bamboo can grow 35 inches in a single day.

6. Corn is the most planted crop in the United States.

7. Wheat is the third-most-produced grain in the world after corn and rice.

8. There are about 40,000 types of rice.

9. Cacti spines can be up to six inches long.

10. The baobab tree is the world's largest succulent, reaching heights of 75 feet.

11. It can live for several thousand years.

12. Dendrochronology is the science of calculating a tree's age by its rings.

13. The first potatoes were cultivated in Peru about 7,000 years ago.

14. Strawberries are the only fruit that bears seeds on the outside.

15. The average strawberry has 200 seeds.

16. Poison ivy produces a skin irritant called urushiol.

17. Moon flowers open in the evening and close before midday.

18. Moon flower vines always coil clockwise.

19. Morning glory blooms open from dawn to midmorning.

20. The California redwood (coast redwood and giant sequoia) are the tallest and largest living organism in the world.

21. A sunflower looks like one large flower, but each head is composed of hundreds of tiny flowers called florets, which ripen to become the seeds.

22. Sunflowers range in height from 1-foot-tall dwarfs to 15-foot-high giants.

23. The lemon tree is the smallest evergreen tree that is native to Asia.

24. The *Rafflesia arnoldii* plant has the largest single bloom of any plant, measuring three feet across.

25. Also known as "meat flower," the parasitic *Rafflesia arnoldii* can hold several gallons of nectar, and its smell has been compared to "buffalo carcass in an advanced stage of decomposition."

26. Aquatic duckweed is the smallest flowering plant on Earth.

27. The plant is only 0.61 millimeter long, and the edible fruit is about the size of a grain of salt.

28. Most coal is formed from ancient plant remains.

29. Stinging nettle leaves contain a mixture of chemicals that it injects into the skin of animals.

30. Dutchman's breeches are named for their flowers, which look like upside-down pants.

31. Thistles have spines on their stems and leaves that keep most animals away from the plant.

32. Bamboo is the world's tallest grass, growing as much as 35.4 inches in a single day.

33. Vanilla comes from a type of orchid.

35. Although tulips are associated with Holland, they are actually not native there; tulips descend mostly from species originating in the Middle East.

36. During the 1600s, tulip bulbs were worth more than gold in Holland.

37. The craze was called tulip mania and crashed the Dutch economy.

38. The Jerusalem artichoke is not really an artichoke.

39. The passionflower gets its name from the complex structure of its flowers, which can be seen as symbols of the crucifixion of Christ.

40. Turkey and Bulgaria are the largest producers of the Damask rose (*Rosa damascena*).

41. Although other plants die if they lose 8 to 12 percent of their water content, the resurrection fern can survive despite losing 97 percent of its water content.

42. The baobab tree can store 1,000 to 120,000 liters of water in its swollen trunk.

43. Spanish moss is neither moss nor Spanish.

44. It's related to the pineapple and has been used to stuff furniture, car seats, and mattresses.

45. The technical term for air plant is epiphyte.

46. These plants get moisture and nutrients from the air.

47. The top of the *Hydnora africana* flower looks like a gaping, fang-filled mouth.

48. It emits a putrid scent to attract dung or carrion beetles.

49. The carnivorous King Monkey Cup traps prey such as scorpions, mice, rats, and birds in pitchers up to 14 inches long and 6 inches wide.

50. It then digests them in a half gallon of enzymatic fluid.

51. Ancient Greeks gave hemlock to prisoners who were condemned to die, including Socrates.

52. Ingesting hemlock will eventually paralyze the nervous system, causing death from lack of oxygen to the brain and heart.

53. California boasts the oldest known living tree—a Bristlecone Pine named "Methuselah."

54. According to tree-ring data, it is 4,853 years old.

43 Amazing Facts About Asia

1. Asia is the most populous continent on Earth, with an estimated 4.7 billion people.

2. Roughly 60 percent of the world's population lives in Asia.

3. Asia is the largest continent with a land area of 17,207,994 square miles.

4. Turkey's capital city of Istanbul is located in both Asia and Europe.

5. Indonesia has the longest coastline of any Asian country at over 33,999 miles.

6. Nepal has the only national flag that is not a rectangle or a square. Its shape evokes the Himalayan peaks.

7. Kyoto, Japan, is home to about 1,660 Buddhist temples and more than 400 Shinto shrines.

8. The Maldives is Asia's smallest country by landmass (115 square miles) and population (425,000).

9. China is the most populous country in Asia and the world with around 1.4 billion people.

10. The overwhelming majority of the Chinese population is found in the eastern half of the country; the west remains sparsely populated.

11. Though ranked first in the world in total population, China's overall density is less than that of many other countries in Asia and Europe.

12. India is the most populous democracy in the world with some 1.3 billion people.

13. Cows are sacred in India.

14. Mount Everest on the China–Nepal border is the tallest peak on Earth at 29,035 feet measured from sea level.

15. The Himalayan jumping spider, which lives on Mount Everest, is the world's highest-dwelling creature.

16. The Dead Sea, between Israel and Jordan, is the world's lowest point at 1,414 feet below sea level.

17. There are more horses than people in the country of Mongolia.

18. At 3,977 miles, the Great Wall of China is the world's longest artificial structure.

19. Borneo is Asia's largest island at 290,321 square miles.

20. The island is shared by three countries: Brunei, Indonesia, and Malaysia.

21. The Forbidden City in Beijing, China, was built with 3.1 million bricks.

22. Bhutan is the world's only Buddhist kingdom.

23. Many of the world's tallest skyscrapers are in Asia, including the 2,717-foot-tall Burj Khalifa building in the United Arab Emirates.

24. The United Arab Emirates is composed of seven sheikh-doms, or emirates: Abu Dhabi, Ajman, Dubai, Fujairah, Ras al Khaimah, Sharjah, and Umm al Quwain.

25. The Pamir and the Tian Shan are among the mountain ranges that cover more than 90 percent of Tajikistan.

26. At 28,251 feet, K2 is Earth's second highest mountain.

27. It was named K2 because it was the second peak measured in the Karakoram Range.

28. Macau, a special administrative region of China, has the world's highest population density with 56,059 people per square mile.

29. Portions of Azerbaijan, Georgia, Kazakhstan, Russia, and Turkey fall within both Europe and Asia, but the larger section of each is in Asia.

30. Although 77 percent of Russia's landmass is in Asia, most Russians live on the European side.

31. Lake Baikal in southern Siberia is Earth's deepest lake at 5,315 feet.

32. It holds approximately one-fifth of the world's fresh surface water.

33. Japan has an average life expectancy of 84.62 years, one of the world's highest.

34. The Son Doong cave in Vietnam's Phong Nha-Ke Bang National Park is the world's largest cave.

35. The mango is the national fruit of India.

36. One of the world's tallest flagpoles is found in Jeddah, Saudi Arabia. It stands 561 feet tall.

37. The national drink of Kyrgyzstan is kumis, made from fermented horse milk.

38. The Philippine archipelago includes over 7,000 islands.

39. There are 12 landlocked countries in Asia: Uzbekistan, Armenia, Turkmenistan, Kyrgyzstan, Afghanistan, Tajikistan, Laos, Mongolia, Kazakhstan, Nepal, Bhutan, and Azerbaijan.

40. Asia is home to more roller coasters than any other continent.

41. Devout Muslims throughout the world turn toward Mecca, Saudi Arabia, in prayer five times each day.

42. The Dome of the Rock, a shrine in Jerusalem that dates to the late 7th century A.D., is the oldest Islamic monument still in existence.

43. The rock over which the shrine was built is sacred to both Muslims and Jews.

40 Cave Facts

1. According to the National Cave and Karst Research Institute, caves are naturally occurring hollow spaces in the ground.

2. They are large enough for a person to enter, and have multiple rooms or passageways to explore.

3. *Speleology* refers to the scientific study and exploration of caves.

4. *Spelunking*, however, refers to the recreational pastime of exploring cave systems.

5. This is also called caving, or potholing, in the United Kingdom.

6. There are several different types of cave, each formed by different natural processes.

7. Lava tubes are formed through volcanic activity.

8. As lava flows down the flanks of volcanoes, the surface of the lava flow, where the molten rock meets the air, cools and hardens first. These sections insulate the lava flowing inside them.

9. When the eruption stops, a hollow tube is left behind.

10. Sometimes, the last remaining bits of lava create stalactites and stalagmites.

11. Just off the southern coast of South Korea, the multi-colored carbonate speleothems of the lava tube cave system in Jeju Volcanic Island attracts many visitors.

12. The air route from Seoul to Jeju Island served over 11 million passengers in 2015.

13. Sea caves occur on nearly every cliffed coast, or coasts where wave action breaks onto cliff faces.

14. Waves are a type of mechanical weathering. They crash into weak zones in sea cliffs and slowly break apart the rock, creating a larger and larger cavity over time.

15. When the same process occurs along lakes, it creates what is known as *littoral caves*.

16. Some sea caves form below sea level, but later emerge above the water due to uplift.

17. Similar processes form eolian caves, which are carved by wind.

18. Small particles of sediments like sand or silt are carried by winds and blasted against cliff faces.

19. A well-known example of an eolian cave is Mesa Verde National Park in Colorado.

20. The homes inside the cave were built by the Ancestral Pueblo people.

21. Glacier caves are created in glaciers because of melting ice and flowing water.

22. The caves tend to be unstable as their shapes and forms change from year to year.

23. Ice caves are different from glacial caves. Ice caves are rock caves that have permanent ice deposits.

24. Solution caves are the most common type of cave.

25. They form in karst, a type of landscape made from soluble rocks like carbonate limestone, dolomite, or marble, or evaporite gypsum, anhydrite, or halite (salt).

26. Rainwater collects carbon dioxide from the atmosphere and soil and forms a weak carbonic acid. The acid seeps through fractures in the ground and dissolves the

surrounding rock. Over time, the cracks become large enough to form caves.

27. Cave formations are called *speleogens* and *speleothems*.

28. Speleogens are created during the formation of the cave.

29. Speleothems are formed later by mineral deposits.

30. Stalactites are icicle-shaped deposits that grow downward from a cave's ceiling.

31. Stalagmites are similar, but grow upward from the floor.

32. Stalactites and stalagmites will occasionally connect, forming columns.

33. Flowstones are sheetlike deposits from which water flows down the walls or along the floors of a cave.

34. Mammoth Cave in Kentucky is the world's largest network of caverns.

35. There are more than 350 miles of underground passages on five different levels. New caves and passageways are still being discovered.

36. Cave of the Crystals is a cave of large selenite crystals underneath the Sierra de Naica Mountain in Chihuahua, Mexico.

37. It was discovered by brothers Juan and Pedro Sanchez.

38. The largest of the cave's crystals is over 37 feet long.

39. In 2018, 12 members of the Wild Boars soccer team and their 25-year-old assistant coach became trapped in Thailand's Tham Luang Nang Non cave.

40. Soon after they entered, heavy rainfall began to flood the cave system, blocking the exit. Monsoons were not expected to start for another month. The boys and their coach were rescued.

46 Facts About Africa

1. Africa is Earth's second largest continent with 11,608,161 square miles.

2. With an area of 226,917 square miles, Madagascar is Africa's largest island.

3. Many kinds of wild guinea fowl are found in Africa. The birds derive their name from a section of Africa's west coast.

4. Two-thirds of Africa lies in the Northern Hemisphere.

5. The world's deepest mine is Harmony Gold's Mponeng gold mine, near Johannesburg, South Africa.

6. It goes nearly two and a half miles deep.

7. Africa has over 25 percent of the world's bird species.

8. The Nile is the world's longest river.

9. It flows for 4,132 miles.

10. The first great civilization in Africa arose 6,000 years ago on the banks of the lower Nile River.

11. Timbuktu, Mali, is home of one of the oldest universities in the world, established in 982 B.C.

12. African elephants, the world's largest living land animal, have ears shaped like the continent of Africa.

13. More than one million western white-bearded wildebeest inhabit the Serengeti Plains and acacia savanna of northwestern Tanzania and adjacent Kenya.

14. Some 1,600 languages are spoken in Africa—more than any other continent.

15. Ethiopia is the only African country with its own indigenous alphabet.

16. The world's largest frog, the Goliath frog, is found in Equatorial Guinea and Cameroon.

17. Libya has no natural rivers.

18. Lake Tanganyika is the world's second deepest lake at 4,820 feet.

19. In 1871, the missing explorer David Livingstone was found by Henry Morton Stanley on the shore of Lake Tanganyika.

20. Victoria Falls, on the border between Zambia and Zimbabwe, is 355 feet high.

21. All of Africa was colonized by foreign powers during the "scramble for Africa" except Ethiopia and Liberia.

22. Zanzibar, an island off the east coast of Africa, is made up entirely of coral.

23. The highest point in Africa is Mount Kilimanjaro in Tanzania at 19,340 feet high.

24. Africa's lowest point is Lake Assal in Djibouti at 512 feet below sea level.

25. The most populous African country is Nigeria with around 206 million people.

26. Africa's longest mountain range is the Atlas Mountains, which stretch 1,553 miles across Morocco, Algeria, and Tunisia.

27. The kingdom of Lesotho is completely encircled by South Africa.

28. Gorée Island, Senegal, was the site of one of the earliest European settlements in Western Africa and long served as an outpost for slave trading.

29. The Valley of the Kings in Egypt was the burial site of almost all the pharaohs of the 18th, 19th, and 20th dynasties.

30. While Egypt is famous for its pyramids, Sudan has double the number of pyramids in Egypt.

31. Lake Victoria is the largest lake in Africa and the second-largest freshwater lake in the world.

32. South Africa is home to the world's oldest mountains.

33. The Barberton Greenstone Belt, also known as the Makhonjwa Mountains, is formed of rocks dating back 3.6 billion years.

34. The World Cup was played for the first time on the African continent in 2010 when South Africa hosted the event.

35. The largest country in Africa is Algeria, which covers 919,595 square miles.

36. The Seychelles is the smallest country in Africa, with an area of just 176 square miles.

37. Deforestation rates in Africa are twice the average for the rest of the world.

38. The vast Etosha Pan, the extremely flat salt pan covering an area of approximately 1,900 square miles, is the largest of its kind in Africa.

39. The largest impact crater on Earth is the Vredefort Crater, near Johannesburg, South Africa, with an estimated diameter of 155 to 186 miles.

40. South Sudan, the newest country in Africa, gained independence from the Republic of Sudan in 2011.

41. Africa has more countries than any other continent with 54.

42. In 1974, scientists discovered the oldest known human ancestor in Ethiopia.

43. The 3.2-million-year-old skeleton was named "Lucy" after the Beatles' song "Lucy in the Sky with Diamonds."

44. The Strait of Gibraltar, which separates Africa (Morocco) from Europe (Spain), is just under nine miles wide at its narrowest point.

45. Monrovia, the capital of Liberia, is the only foreign capital city named after a U.S. president.

46. James Monroe was president when the city was founded in 1822 as a refuge for freed slaves.

10 Volcanic Facts

1. There are some 1,900 active volcanoes on Earth.

2. The Ring of Fire is a seismically active belt of volcanoes and tectonic plate boundaries that stretches 24,900 miles around the Pacific Ocean.

3. About 75 percent of the world's volcanoes occur within the Ring of Fire.

4. The word "volcano" comes from the name Vulcan, the Roman god of fire.

5. Mauna Loa in Hawaii is the world's largest active volcano, rising approximately 30,000 feet from the base at the bottom of the ocean to its peak.

6. Molten rock is called magma when it's still under the earth, and lava when it comes spewing out of a volcano.

7. Fresh lava can reach temperatures as high as 2,200 °F.

8. Nevado Ojos del Salado on the Argentina–Chile border is the highest active volcano in the world at 22,615 feet above sea level.

9. The most famous eruption of Mount Vesuvius, in southern Italy, occurred in A.D. 79, when lava and ashes buried the towns of Pompeii and Herculaneum.

10. The Hawaiian Islands are composed of shield volcanoes that have built up from the seafloor to the surface.

Chapter 2

HISTORY

42 Facts About Ancient Civilizations

1. Mesopotamia, located between the Tigris and Euphrates rivers in modern-day Iraq, has been home to many great civilizations: Sumer, Babylonia, and Assyria.

2. Mesopotamia means "between two rivers" in Greek.

3. Most Western scholars agree that the Sumerian civilization in Mesopotamia was the world's first civilization.

4. Ancient Sumer is believed to have begun around 4000 B.C.

5. Archaeological evidence suggests that "pre-civilized" cultures lived in the Tigris and Euphrates river valleys long before the emergence of Sumer.

6. The great city of Ur, associated with Sumer, is possibly the world's first city.

7. The Harappan civilization developed in the Indus River Valley in modern-day Pakistan and India, beginning around 3500 B.C.

8. However, it is clear that agricultural communities had inhabited the Indus Valley area since at least 9000 B.C.

9. The Harappan civilization (circa 3300–1600 B.C.) boasted a network of earthenware pipes that carried water from people's homes into municipal drains and cesspools.

10. Nearly every home in the cities of Harappa and Mohenjo-Daro had a toilet connected to a sophisticated sewage system.

11. Located in Africa's Nile Valley, ancient Egypt is generally cited as beginning in 3200 B.C., though agricultural

societies had settled in the Nile River Valley since the tenth millennium B.C.

12. Ancient Egyptian doctors had specialties. Dentists were known as "doctors of the tooth," while the term for proctologists literally translates to "shepherd of the anus."

13. The Elamite kingdom in modern-day Iran began around 2700 B.C., though recent evidence suggests that a city existed in this area at a far earlier date—perhaps early enough to rival Sumer.

14. Meanwhile, the ancient Chinese civilizations, located in the Yangtze and Yellow river valleys, are said to have begun around 2200 B.C.

15. The earliest paved roads go back all the way to 4000 B.C., in the Mesopotamian cities of Ur and Babylon.

16. In the southern region of Sumer, brick makers would set their bricks in place with a substance called bitumen, also known as asphalt.

17. The Akkadians formed the first united empire where the city-states of Sumer were united under one ruler.

18. The earliest writing, cuneiform, was inscribed on clay tablets and dates to around 3200 B.C.

19. Over a period of 3,000 years, it was used to record about 15 different languages, including Sumerian, Akkadian, Babylonian, and Assyrian.

20. Ancient Mesopotamians built terraced pyramids called ziggurats to honor the gods.

21. The ancient city of Babylon, located on the modern-day site of Al Hillah, Iraq, became the most powerful city in Mesopotamia.

22. The ancient Babylonians worshipped a half-human/half-fish creature named Oannes who gave them the gift of civilization.

23. The saying "an eye for an eye" comes from the Code of Hammurabi—a system of 282 laws created by Babylonian King Hammurabi in 1750 B.C.

24. Babylon was said to have been surrounded by walls 100 feet wide, 300 feet high, and 56 miles in length.

25. The walls were built with some 15 million bricks.

26. According to legend, King Nebuchadnezzar II built the Hanging Gardens of Babylon around 600 B.C. as a present for his wife.

27. There is some speculation over whether or not the Hanging Gardens of Babylon, one of the seven wonders of the ancient world, ever existed.

28. The oldest known furnace for copper smelting was discovered at Timna in southern Israel, and dated to approximately 4200 B.C.

29. It was simply a small hole in the ground that could be covered with a flat rock.

30. The Sumerians are often credited with inventing the wheel.

31. Ancient Sumerian kings and queens played board games.

32. The ancient Maya developed a picture-based written language around 300 B.C.

33. The Chavín, the earliest ancestors of the Inca, arrived in modern-day Peru around 1200 B.C.

34. Australian Aborigines, the world's oldest living culture, have existed for at least 50,000 years.

35. The earliest writing in ancient China dates to around 1200 B.C. It was carved into animal bones.

36. More than 10,000 individual pieces of graffiti have been found in the ancient Roman city of Pompeii, on

surfaces ranging from bathroom walls to the outer walls of private mansions.

37. Ancient Romans ran cold aqueduct water in pipes through their houses in an early form of air conditioning.

38. Between 750 and 550 B.C., the ancient Greeks established 200 colonies throughout the Mediterranean.

39. Ireland's Newgrange, an ancient Celtic site, is 60 years older than Egypt's Giza Pyramids and 1,000 years older than England's Stonehenge.

40. Ancient Persian engineers built a type of evaporative cooler called "Yakhchāl" that could store ice even in the middle of summer.

41. According to the ancient Greek historian Herodotus, ancient Persians would get drunk after making an important decision to see if they felt the same way about it when intoxicated.

42. The earliest coins were used in the ancient kingdom of Lydia (modern-day Turkey) in the seventh century B.C.

11 of History's Shortest Wars

1. **Anglo-Zanzibar War** (9:02–9:40 A.M., Aug. 27, 1896, Great Britain vs. Zanzibar): After Khalid bin Barghash acceded to the sultanate of Zanzibar (an island off modern Tanzania) without prior British permission, the Royal Navy gave Zanzibar a taste of British anger. Bin Barghash tapped out after just 38 minutes of shelling in what is history's shortest recorded war.

2. **Spanish-American War** (Apr. 25–Aug. 12, 1898, U.S. vs. Spain): Spain once had an empire, some of which was near Florida. After months of tension, the USS *Maine* mysteriously blew up in Havana Harbor. Though no one knew why it exploded, the U.S. declared war anyway. A few months later, Spain had lost Cuba, Guam, the Philippines, and Puerto Rico.

3. **Nazi-Polish War** (Sep. 1–Oct. 6, 1939, Nazi Germany and Soviet Union vs. Poland): After Russian and German negotiators signed a secret agreement in August dividing Poland, the Nazis invaded in vicious armored thrusts with heavy air attacks. Polish forces fought with valor, but their strategic position was impossible. Russian troops entered from the east on September 17, and Poland became the first European nation conquered in World War II.

4. **Nazi-Danish War** (4:15–9:20 A.M., Apr. 9, 1940, Nazi Germany vs. Denmark): Arguably the biggest mismatch of World War II (unless one counts Germany's invasion of Luxembourg). Sixteen Danish soldiers died before the Danish government ordered the resistance to cease.

5. **Suez/Sinai War** (Oct. 29–Nov. 6, 1956, Israel, Britain, and France vs. Egypt): The Egyptians decided to nationalize the Suez Canal, which seems logical today given that the Suez is entirely in Egypt. British and French companies operating the canal didn't agree. The Israelis invaded by land, the British and French by air and sea. The invaders won a complete military victory, but withdrew under international pressure.

6. **Six-Day War** (June 5–10, 1967, Israel vs. Egypt, Syria, and Jordan): Israelis launched a sneak attack on Egyptians, destroying the Egyptian air force on its airfields and sending the Egyptians reeling back toward the Suez Canal. Jordanians attacked the Israelis and immediately regretted it. The Israelis attacked Syria and seized the Golan Heights.

7. **Yom Kippur War** (Oct. 6–25, 1973, Egypt and Syria vs. Israel): Egyptians and Syrians, still annoyed and embarrassed over the Six-Day War, attacked Israelis on a national religious holiday. Israeli forces were caught napping at first but soon regained the upper hand—they struck within artillery range of Damascus and crossed

the Suez Canal. The United Nations' ceasefire came as a major relief to all involved.

8. **Soccer War** (July 15–19, 1969, El Salvador vs. Honduras): Immigration was the core issue, specifically the forced expulsion of some 60,000 Salvadoran illegal immigrants from Honduras. When a soccer series between the two nations fueled tensions, each managed to insult the other enough to start a bloody yet inconclusive war.

9. **Falklands War** (Mar. 19–June 14, 1982, Argentina vs. U.K): Argentina has long claimed the Falkland Islands as Las Islas Malvinas. In 1982, Argentina decided to enforce this claim by invading the Falklands and South Georgia. Although the Argentines had a surprise for the Royal Navy in the form of air-launched antiship missiles, the battle for the islands went heavily against Argentina. Its survivors, including most of its marines, were shipped home minus their weaponry.

10. **Invasion of Grenada** (Oct. 25 to mid-Dec., 1983, U.S. vs. Grenada and Cuba): Concerned about a recent Marxist takeover on the Caribbean island of Grenada, elite U.S. forces invaded by air and sea. Some 1,000 Americans (including 600 medical students) were in danger in Grenada; after the relatively recent Iran hostage crisis, the lives of U.S. citizens overseas were a powerful talking point in domestic politics.

11. **First Gulf War** (Jan. 16–Mar. 3, 1991, Allies vs. Iraq): Saddam Hussein misjudged the world's tolerance for military adventurism (at least involving oil-rich Western-friendly Arab emirates) by sending his oversized military into Kuwait. President George H.W. Bush led a diverse world coalition that deployed into Saudi Arabia and started bombing the Iraqi military. Most remarkable: The primary land conflict of the war lasted just 100 hours, February 23–27.

50 Facts About Ancient Egypt

1. Ancient Egypt was one of the most advanced and stable civilizations in human history, lasting for more than 3,000 years.

2. Ancient Egypt lasted from around 3100 B.C. to 30 B.C., when it was conquered by the Romans.

3. Ancient Egyptians called their picture-based written language *medu netjer*, which means "words of god."

4. The ancient Greeks renamed the writing system hieroglyphs, or "sacred carvings."

5. The names of important people were written inside an oval called a cartouche.

6. As the breadbasket of the eastern Mediterranean and a gateway to African trade, geography guaranteed that major powers would covet Egypt.

7. Ramesses II, also known as Ramesses the Great, made the world's oldest known peace treaty after a long war against the Hittites ended in a draw in 1274 B.C.

8. Some of Egypt's greatest rulers and heroes were women.

9. The best example is Hatshepsut, who reigned as Pharaoh of Egypt rather than as queen.

10. Ancient reports during a military campaign in Nubia tell of the homage Hatshepsut received from rebels defeated on the battlefield.

11. Pharaohs wore a cobra symbol on their crowns, which was believed to spit fire at the pharaoh's enemies.

12. Both men and women wore makeup in ancient Egypt.

13. Eyeliner was drawn on with kohl, a black powder.

14. Nefertiti was the wife of Pharaoh Akhenaten. She is believed to have ruled beside him for 14 years.

15. Some scholars believe Nefertiti ruled briefly as pharaoh after her husband died.

16. Nefertiti was made famous by her bust, one of the most copied works of ancient Egyptian art.

17. Shaving away all body hair—most notably the eye-brows—was part of an elaborate daily purification ritual that was practiced by the pharaoh and his priests.

18. The ancient Egyptians believed that health, good crops, victory, and prosperity depended on keeping their gods happy.

19. Egyptian religion revolved around dozens of gods, often part animal.

20. Shaving the eyebrows was also a sign of mourning, even among commoners.

21. The Greek historian Herodotus, who traveled and wrote in the fifth century B.C., said that everyone in an ancient Egyptian household would shave his or her eyebrows following the natural death of a pet cat.

22. For dogs, Herodotus reported, the household members would shave their heads and all of their body hair as well.

23. Chairs in ancient Egypt had legs shaped like animal limbs.

24. Ancient Egyptian women enjoyed equal privileges with men on many fronts, including the right to buy, sell, and inherit property; to marry and divorce; and to practice an occupation outside the home.

25. Legal rights and social privileges varied by social class rather than by gender. In other words, women and men in the same social class enjoyed fairly equal rights.

26. Queen Ahhotep, of the early 18th Dynasty, was granted Egypt's highest military decoration—the Order of the Fly—at least three times for saving Egypt during the wars of liberation against the Hyksos invaders from the north.

27. Egyptian pharaohs often married their siblings because it was believed that pharaohs were gods on Earth and thus could marry only other gods.

28. The entire Egyptian Empire depended on Nile floods. Ancient Egyptian famers built ditches and low walls to trap flood water from the Nile.

29. Ancient Egyptians brewed and drank beer. It was actually a source of nutrition, motivation, and sometimes payment for work.

30. Ancient Egyptians lined linen bandages with honey to get them to stick to skin.

31. Tutankhamun became pharaoh in 1336 B.C. at the age of nine.

32. He ruled until 1327 B.C. when he died suddenly at the age of 18.

33. Tutankhamun's tomb was discovered in 1922.

34. Graverobbers had looted the tombs of many other pharaohs, but Tut's tomb remained undisturbed for nearly 3,000 years.

35. It took archaeologists 10 years to catalog the contents of Tut's tomb.

36. The earliest step pyramid dates to around 2630 B.C.

37. At Saqqara, Egypt, the architect Imhotep built mastaba upon mastaba, fashioning a 200-foot-high pyramid as a mausoleum for Pharaoh Djoser.

38. Over the next thousand years or so, Egyptian engineers built several pyramid complexes along the Nile's west bank. The most famous and popular today are those at Giza, but dozens survive.

39. We used to believe pyramids were built under the lash, but modern scholars doubt this.

40. Egyptians believed that after a pharaoh died, he would reach the heavens via sunbeams.

41. Egyptians also believed that the pyramid shape would help the pharaoh on this journey.

42. Once the pharaoh reached the heavens, he would become Osiris, the god of the dead.

43. To adequately perform the duties of Osiris, the pharaoh would need a well-preserved body and organs.

44. The Egyptians feared that if they did not prepare the pharaoh's body for the afterlife, disaster would fall upon Egypt.

45. Ancient Egyptian children played a game similar to hockey with sticks made from palm branches and pucks made of leather pouches stuffed with papyrus.

46. In 332 B.C., Alexander the Great of Greece conquered Egypt.

47. Pharaohs in the Ptolemaic Dynasty were Greeks descended from Ptolemy, one of Alexander's generals, so Greek was the dominant language of their court.

48. Pharaoh Ptolemy Soter began building the world's first lighthouse in the harbor of Alexandria, Egypt, in 290 B.C.

49. Cleopatra VII was the last pharaoh. She and Mark Antony killed themselves.

50. Egypt became part of the Roman Empire in 30 B.C., ending the age of the pharaohs.

39 Marvelous Mummy Facts

1. Ancient Egyptians might be history's most famous embalmers, but they weren't the first.

2. In northern Chile and southern Peru, modern researchers have found hundreds of pre-Inca mummies (roughly 5000–2000 B.C.) from the Chinchorro culture.

3. The Chinchorros mummified all walks of life: rich, poor, elderly, didn't matter.

4. One of the oldest known instances of deliberate mummification comes from a rock shelter now called Uan Muhuggiag in southern Libya: the well-preserved mummy of a young boy.

5. Radiocarbon dating determined the age of the Tashwinat Mummy discovered at Uan Muhuggiag to be approximately 5,600 years old.

6. According to Egyptian lore, the god Osiris was the very first mummy.

7. The earliest known Egyptian mummy dates to around 3300 B.C.

8. As Egyptian civilization advanced, professionals formalized and refined the process of mummification.

9. *Natron*, a mixture of sodium salts abundant along the Nile, made a big difference.

10. If you extracted the guts and brains from a corpse, then dried it out it in natron for a couple of months, the remains would keep for a long time.

11. During mummification, ancient Egyptians removed all internal organs except the heart.

12. The heart was left in the body because the Egyptians believed that—as the seat of intelligence and emotions—the person would need his heart in the afterlife.

13. The canopic jars used to store a mummy's internal organs are named after the local god Canopus. Canopus is also a town in the Nile delta region.

14. The entire mummification process took 40 to 70 days in ancient Egypt.

15. Many people were employed in mummification, including the embalmer, cutters, priests, scribes, servants, and other workers.

16. In ancient Egypt, chief embalmers often wore a mask of Anubis, the god of mummification and the afterlife.

17. Only the rich in ancient Egyptian society could afford to be mummified.

18. Egyptians used vast amounts of linen to mummify a body.

19. There was enough linen on one mummy from the 11th dynasty to cover three tennis courts.

20. Ancient Egyptians mummified animals, including cats, jackals, baboons, horses, birds, gerbils, fish, snakes, crocodiles, bulls, and hippos.

21. The most popular mummy in the world is likely Vladimir Lenin.

22. Millions of people have visited his mummy in Moscow, Russia.

23. Mummies have been found on every continent except Antarctica.

24. The Spirit Cave Mummy is the oldest mummy found in North America.

25. Unearthed in 1940 in Nevada, the naturally preserved 9,400-year-old was shrouded in woven reed mats and a rabbit-skin blanket.

26. The oldest well-preserved mummy in Europe is Otzi the Iceman, a natural mummy from 3300 B.C. found in the Alps.

27. Another famous mummy is Lady Dai (Xin Zhui), who died around 168 B.C.

28. Lady Dai's are some of the best preserved human remains ever discovered in China.

29. She was buried in an immense tomb at Mawangdui in China and unearthed in 1968.

30. Xin Zhui was married to Li Cang, the Marquis of Dai and Chancellor of Changsha Kingdom in ancient China.

31. The contents of Xin Zhui's tomb revealed much previously unknown information about life in the Han dynasty.

32. Perhaps the most famous mummy in history is that of Egypt's boy king Tutankhamun.

33. British archaeologist Howard Carter and his financier friend Lord George Carnarvon opened Tutankhamun's burial chamber in early 1923.

34. Two months later, Carnarvon was dead. His death is often offered as proof of the "Curse of King Tut."

35. The inner coffin of Tutankhamun is made of solid gold weighing 296 pounds.

36. Pharaoh Ramses II was the first mummy to receive a passport (needed for travel). His listed occupation is 'king.'

37. Puruchuco, a site near Lima, Peru, contained more than 2,200 mummies.

38. In Victorian England, mummy unwrapping parties were popular. The host would buy one, then the guests would unwrap it.

39. People in Victorian times also used ground up mummies for medicinal purposes.

Egyptian Mummification in 7 Steps

1. Wash and ritually purify body.

2. Remove intestines, liver, stomach, and lungs; embalm them with natron (soda ash) and place in jars.

3. Stuff body cavity with natron.

4. Remove brain through nose using a hook; throw brain away.

5. Cover body with natron and place on embalming table for 40 days.

6. Wrap 20 layers of linen around body, gluing linen strips together with resin.

7. Place mummy in protective sarcophagus, and add another layer of wrapping.

10 Military History Facts

1. The oldest military medal? Probably the Gold of Valor. It was awarded by the Egyptian pharaohs around 1500 B.C.

2. The Hundred Years War lasted 116 years.

3. During his invasion of England in 1014, King Olaf's fleet of Viking ships managed to pull down London's wooden Thames River bridge. (Hence, the children's song "London Bridge Is Falling Down.")

4. The longest-running mercenary contract belongs to the world's smallest standing army—the Vatican's Swiss Guards, a 100-man company first hired by Pope Julius II in 1506.

5. Sailing into space: Pieces of Germany's High Seas Fleet, scuttled off Scotland at the end of World War I, have been used to build deep-space probes.

6. What's with General George Patton's ivory-handled revolvers? He started carrying revolvers in 1916 after he nearly blew his own leg off with the Army's newfangled automatic pistol.

7. The U.S. Marines' first land battle on foreign soil was in Derna, Libya, in 1805. During this battle, 600 Marines stormed the city to rescue the crew of the USS *Philadelphia* from pirates.

8. In 1940, Brigadier General Benjamin Oliver Davis became the first African American general in the U.S. Armed Forces.

9. Britain's early 20th-century super-weapon, the battle-ship H.M.S. *Dreadnought*, was fitted with ultramodern weapons, so her builders naturally omitted the ancient ram from her design. Her only kill was a submarine, which she sunk by ramming it.

10. Eighty percent of Soviet males born in 1923 didn't survive World War II.

50 Facts About Ancient Rome

1. Ancient Romans ruled over one of the largest and richest empires in history.

2. Rome was founded in the ninth century B.C. as a small village in central Italy. Over the centuries it expanded its territory.

3. At its largest, the Roman Empire covered almost 2 million square miles.

4. The Roman Empire was most powerful between the third century B.C. and the fifth century A.D., when their armies conquered most of Europe, North Africa, and the Middle East.

5. In its early days, Rome was ruled by kings.

6. In 509 B.C., the kingdom became a republic, ruled by a council called the Senate.

7. The highest position in the Roman Republic was the consul.

8. There were two consuls at the same time to make sure that one didn't become too powerful.

9. Pirates captured a twentysomething Julius Caesar at sea. After his release, Caesar raised a fleet, hunted the pirates down, and crucified them all.

10. As its territory grew, army generals threatened to take power away from the Senate.

11. In 45 B.C., Julius Caesar took control of Rome and became its dictator.

12. Just a year later, Caesar was assassinated on March 15 (the Ides of March).

13. Caesar was stabbed 23 times by a group of senators who believed he was undermining the republic.

14. Later, his adopted son, Augustus, became the first emperor of Rome.

15. As a young man Augustus was known as Octavian.

16. When Julius Caesar was assassinated, Octavian led his armies to victory in the battle for power over Rome.

17. Octavian proclaimed himself "First Citizen" and took the name Augustus.

18. Emperors ruled Rome from 27 B.C. until the end of the empire.

19. The word emperor comes from the Latin *imperator*, meaning "military commander."

20. Emperor Augustus attempted to reform what he considered loose Roman morals.

21. In a bit of irony, Augustus then had to banish his daughter Julia for committing adultery.

22. Constantine I brought the Roman Empire together again in A.D. 324 after years of turmoil.

23. Constantine was the first Christian emperor of Rome and he issued laws protecting Christians from persecution.

24. A haruspex was a Roman priest who read messages from the gods by cutting open animals to look at their livers.

25. Before going on a long journey, Romans would bargain with the gods for safety by promising some act of devotion upon arrival.

26. If the gods finked and the trip went sour, the pious supplicant would kick the god's statue on his or her return.

27. There were only 22 letters in the Roman alphabet. J was written as I, U was written as V, and W and Y did not exist.

28. Many European languages, including English, still use the Roman alphabet today.

29. Every male Roman child remained under his father's authority until that father's death. *Pater* could sentence *filum* to death; disinherit him; block him from legal action on his own behalf; even prevent him from borrowing money.

30. The Romans were the first people to use concrete, which allowed them to build massive structures quickly and cheaply.

31. In Rome, if someone got behind on a debt, the creditor might hire a *convicium* (escort) to serenade the deadbeat with ridicule.

32. Roads built by the ancient Romans are still used today.

33. Mark Antony's wife Fulvia (77–40 B.C.) was a capable general who led eight legions in a rebellion against land confiscation, fighting to the bitter end before going into exile.

34. Ancient Romans wore *soccus*, the predecessors of socks.

35. A fourth-century survey of Rome counted more than 45,000 tenements, nearly 1,800 villas, 850 bathhouses, 1,300 swimming pools, and a dozen each of libraries and toilet complexes. Of course, by then, Rome had receded from its peak glories.

36. The Circus Maximus, a huge stadium built for chariot races, could accommodate over 150,000 spectators.

37. In ancient Rome, urine was as much a commodity as a waste product. The ammonia in pee was useful for bleaching togas and tunics.

38. Emperor Caligula (reigned A.D. 37–41) led a military expedition to the North Sea, where he ordered his soldiers to gather seashells, then claimed he defeated the ocean.

39. Caligula was eventually assassinated by his bodyguards.

40. Claudius (reigned A.D. 41–54) was partially deaf and walked with a limp due to childhood illness.

41. His family ostracized him and kept Claudius out of power.

42. When Caligula was assassinated, Claudius was one of the few survivors from the emperor's family.

43. His infirmity probably saved him from being seen as a serious threat and therefore murdered.

44. Claudius was named emperor and reigned for 13 years.

45. The Bay of Neapolis (now Naples) was Rome's Padre Island or Cabo San Lucas. It was where one went for intercourse and intoxication. Not that genteel Roman society was shy about either.

46. Nero (reigned A.D. 54–68) was a bloodthirsty emperor who had his mother and wife killed. He probably didn't start the Great Fire of Rome though.

47. Emperor Hadrian (reigned A.D. 117–138) realized the Roman Empire was too large to defend and ordered his armies to withdraw to borders that were easier to protect.

48. One of these borders was Hadrian's Wall, a 73-mile line of forts and barriers that separated Roman Britain from the unconquered lands to the north.

49. The fall of the Western Roman Empire in A.D. 476 is considered the start of the "Dark Ages" (or Middle Ages) in Europe.

50. The eastern part of the empire remained a power through the Middle Ages until its fall in A.D. 1453.

25 Colosseum Facts

1. The Colosseum was one of the greatest buildings in the ancient city of Rome, at the heart of the powerful Roman Empire.

2. Construction began under Emperor Vespasian in A.D. 72 and was completed in A.D. 80 by his successor and heir, Titus.

3. Romans celebrated in A.D. 80 with 100 days of inaugural games that included the slaying of 9,000 wild animals, noonday executions, and gladiatorial brawls that usually ended in death.

4. Travertine stone was used to make up much of the exterior of the elliptical building.

5. The arena floor was made of wooden planks supported by the brick walls of the cellars underneath.

6. Sand (which was good for absorbing blood) was scattered across the arena floor.

7. Wild animals that fought in the Colosseum were kept in cellars underneath the floor.

8. When it was time for the animals to appear, they were pushed into cage elevators and winched upward.

9. There were 36 trapdoors in the floor of the Colosseum.

10. The Colosseum had 80 entrances and could hold more than 50,000 roaring Romans thirsty for a gruesome battle.

11. Seats near the front were for the rich and powerful, while regular people sat farther back.

12. The emperor had his own box.

13. There were more than 20 different types of gladiators, each with different weapons and armor.

14. Most gladiators were slaves or criminals who trained in special schools.

15. *Bestiarii* fought wild animals to entertain the public.

16. Wild beast fights usually took place just prior to fights between gladiators.

17. Exotic animals were brought to the Colosseum from all over the Roman Empire.

18. A *secutor* (or chaser) was a close-combat fighter.

19. His traditional opponent was the *retiarius*.

20. The most lightly armed gladiator, the *retiarius* (net man) had to rely on speed and agility to escape his opponent.

21. The retiarius carried a three-pronged spear (trident) and a weighted net.

22. Roman audiences loved watching the retiarius fight slower, more heavily armed opponents.

23. The *hoplomachus* was one of the most heavily armored gladiators, with a large helmet and thick coverings over his arms and legs.

24. His fighting style was based on the warriors of ancient Greece.

25. The Colosseum measures 612 feet long, 151 feet wide, and 157 feet high, making it the world's largest amphitheater.

14 Greek and Roman Gods

Greek God—Job Title—Roman God

1. Aphrodite—Goddess of love and beauty—Venus

2. Apollo—God of beauty, poetry, music—Apollo

3. Ares—God of war—Mars

4. Artemis—Goddess of the hunt and moon—Diana

5. Athena—Goddess of war and wisdom—Minerva

6. Demeter—Goddess of agriculture—Ceres

7. Dionysus—God of wine—Bacchus

8. Hades—God of the dead and the Underworld—Pluto

9. Hephaestus—God of fire and crafts—Vulcan

10. Hera—Goddess of marriage (Queen of the gods)—Juno

11. Hermes—Messenger of the gods—Mercury

12. Hestia—Goddess of the hearth—Vesta

13. Poseidon—God of the sea, earthquakes, and horses—Neptune

14. Zeus—Supreme god (King of the gods)—Jupiter

48 Amazing Ancient Greece Facts

1. Ancient Greece is divided into three main periods: the Archaic Period (800–480 B.C.), the Classical Period (480–323 B.C.), and the Hellenistic Period (323–146 B.C.).

2. Greek poet Homer wrote the *Odyssey* and the *Iliad* during the Archaic Period.

3. Ancient Greece was ruled by city-states, which sometimes fought each other and sometimes joined forces against invaders.

4. A Greek city-state, which the Greek themselves called a *polis*, was a city and the rural lands outside its walls.

5. The two most powerful city-states were Athens and Sparta.

6. Athenians invented democracy around the fifth century B.C.

7. The word "democracy" comes from two Greek words that mean people (*demos*) and rule (*kratos*).

8. Women, children, slaves, and Greeks not born to Athenian parents couldn't vote.

9. Slaves made up approximately one-third of the population of some ancient Greek cities.

10. Sparta had a strong military culture. Boys started harsh military training at a young age.

11. Citizens of Sparta voted by pounding on their shields.

12. Ancient Greeks invented philosophy, which means love of wisdom.

13. Famous Greek philosophers include Aristotle, Socrates, and Plato.

14. Alexander the Great, one of history's most successful military commanders, expanded the ancient Greek empire surrounding the Mediterranean all the way east to India.

15. The Hellenistic Period began when Alexander the Great died and ended when Rome conquered Greece.

16. The Romans copied much of the Greek culture including their gods, architecture, language, music, and even how they ate!

17. Ancient Greeks often ate dinner while lying on their sides.

18. The first two letters of the Greek alphabet—alpha and beta—made up the word alphabet.

19. Their alphabet was the first with vowels.

20. Spiked dog collars were invented in ancient Greece and were originally designed to protect the dog's throat from wolf attacks.

21. Ancient Greeks loved the theater, but only male actors were allowed to perform.

22. In Greek mythology, a sphinx had a woman's head and a lion's body.

23. Ancient Greeks believed their gods lived atop Mount Olympus.

24. They were ruled by Zeus, the king of the gods.

25. Draco was an elected Athenian leader who established a written code of laws.

26. Many punishments were particularly harsh. The adjective "draconian" is now commonly used to describe similarly unforgiving rules or laws.

27. The very first Olympic Games were held in 776 B.C. in the Greek city of Olympia.

28. The Statue of Zeus at Olympia was constructed by Phidias around 435 B.C.

29. Standing 40 feet high with a 20-foot base, the Statue of Zeus was one of the seven wonders of the world.

30. The statue depicted a seated Zeus holding a golden figure of the goddess of victory in one hand and a staff topped with an eagle in the other.

31. After the old gods were outlawed by Christian emperor Theodosius, the statue was taken as a prize to Constantinople, where it was destroyed in a fire around A.D. 462.

32. Greek hero Pheidippides ran from Athens to Sparta to request help in the Battle of Marathon against the Persians.

33. After the Greeks won, he ran 25 miles from Marathon to Athens to announce the victory. This is where the word "marathon" originated.

34. Ancient Greeks wore draped tunics called chitons.

35. Alexander the Great never lost a battle.

36. Greek scientist Hippocrates is called the Father of Western Medicine.

37. Doctors still take the Hippocratic Oath today.

38. Coins in ancient Greece were decorated with images of bees.

39. Ancient Greeks didn't call their land Greece. Its official name was Hellenic Republic, but Greeks have called it Hellas or Hellada.

40. The English word "Greece" comes from the Latin word *Graecia*, meaning "the land of the Greeks."

41. The Greeks used a trick wooden horse to finally win the war against the Trojans.

42. Ancient Greek warriors were well organized and heavily armored.

43. They fought in a rectangular group called a phalanx, covering themselves with shields for protection.

44. Jury trials in ancient Greece had as many as 500 jurors.

45. For breakfast, Greeks ate bread dipped in wine.

46. Ancient Greeks drank wine, usually mixed with water, with meals.

47. Sparta was the only Greek city-state to mandate public education for girls.

48. Most of those splendid Greek statues you see in off-white marble used to be painted in realistic colors. We know because tiny bits of paint remain embedded in nooks and crannies.

The 9 Muses

The nine muses were Greek goddesses (the daughters of Zeus and Mnemosyne) who ruled over the arts and sciences and offered help and inspiration to mortals.

1. Calliope—Muse of epic poetry

2. Clio—Muse of history

3. Erato—Muse of love poetry

4. Euterpe—Muse of music

5. Melpomene—Muse of tragedy

6. Polyhymnia—Muse of sacred poetry or mime

7. Terpsichore—Muse of dance

8. Thalia—Muse of comedy

9. Urania—Muse of astronomy

37 Facts About the Maya

1. The Maya or Mayan peoples made their home in modern-day Mexico and Central America.

2. Mayan culture was well established by 1000 B.C. and lasted until A.D. 1697.

3. The Maya civilization consisted of a large number of cities.

4. The cities shared a common culture and religion, but each city had its own noble ruler.

5. Mayan kings frequently fought with each other over tribute (gifts) and prisoners to sacrifice to the gods.

6. The ancient Maya developed a picture-based written language around 300 B.C.

7. Mayan weapons were made from volcanic rock called obsidian.

8. Ancient Maya nobles pierced their skin with spines from stingrays as part of a blood sacrifice made to the gods.

9. The Maya cultivated stingless bees for honey.

10. There was never a single Mayan empire, but rather a widespread, interconnected civilization.

11. The Maya believed their rulers could communicate with the gods through ritual bloodletting and human sacrifice.

12. Self-sacrifice was also common among the Maya.

13. Crossed eyes, flat foreheads, and big noses were attractive features to the Maya.

14. Mothers tried to flatten their baby's foreheads with boards and purposely make their children cross-eyed since these were features of nobility.

15. Mayan noblewomen filed their teeth into sharp points.

16. Mayans grew a lot of maize (yellow corn), which was a staple in their diet, along with avocados, tomatoes, chili peppers, beans, papayas, pineapples, and cacao.

17. Corn was made into a kind of porridge called *atole*.

18. The Maya were among the first cultures to use saunas.

19. Mayans played a ballgame on a court in which the losers were sacrificed to the gods.

20. Mayan kings were thought to become gods when they died.

21. The Maya were expert mathematicians and astronomers.

22. Mayans followed a 52-year Calendar Round, which resulted in two calendar cycles: the Haab and the Tzolkin.

23. The Haab was made up of 365 days organized into 18 months of 20 days, with 5 unlucky days added at the end.

24. In the Tzolkin or Sacred Round, 20 day names were combined with 13 numbers to give a year 260 days.

25. The two cycles reach the same point after 52 years.

26. For periods longer than 52 years, the Maya used a separate system called the Long Count.

27. The Maya were the first civilization to use the number zero as a place holder, but they were not the first to use it in mathematics.

28. Jade was the most precious material to the Maya.

29. It was associated with water, the life-giving fluid, and with the color of the corn plant, their staple food.

30. Both men and women wore jewelry, including ear spools, necklaces, and bracelets.

31. The Maya used human hair for sutures.

32. Although some Mayan cities continued to thrive until the 16th century, the Mayan civilization began to decline after A.D. 800.

33. During the 9th century A.D., cities in the central Maya region began to collapse, likely due to a combination of causes, including endemic inter warfare, overpopulation, severe environmental degradation, and drought.

34. During this period, known as the Terminal Classic, the northern cities of Chichén Itzá and Uxmal showed increased activity.

35. Many cities such as Tikal (in modern-day Guatemala) were swallowed up by the rain forest and not rediscovered until modern times.

36. Tikal was one of the biggest Mayan cities, with as many as 60,000 people living there.

37. The last surviving Mayan city was Tayasal, which existed until 1696 when the Spanish conquered it.

11 Ancient Cities: Then and Now

1. **Memphis** (now the ruins of Memphis, Egypt): By 3100 B.C., this Pharaonic capital bustled with an estimated 30,000 people. Today it has none—but modern Cairo, 12 miles north, is home to an estimated 21,750,000 people.

2. **Ur** (now the ruins of Ur, Iraq): Sumer's great ancient city once stood near the Euphrates with a peak population of 65,000 around 2030 B.C. Ur now has a population of zero.

3. **Alexandria** (now El-Iskandariyah, Egypt): Built on an ancient Egyptian village site near the Nile Delta's west end, Alexander the Great's city once held a tremendous library. In its heyday, it may have held 250,000 people; today an estimated 5.4 million people call it home.

4. **Babylon** (now the ruins of Babylon, Iraq): Babylon may have twice been the largest city in the world, in about 1700 B.C. and 500 B.C.—perhaps with up to 200,000 people in the latter case. Now, it's windblown dust and faded splendor.

5. **Athens** (Greece): In classical times, this powerful city-state stood miles from the coast but was never a big place—something like 30,000 residents during the 300s B.C. It now reaches the sea with about 3,154,000 residents.

6. **Rome** (Italy): With the rise of its empire, ancient Rome became a city of more than 500,000 and the center of Western civilization. Though that mantle moved on to other cities, Rome now has around 4.3 million people in the metro area.

7. **Xi'an** (China): This longtime dynastic capital, famed for its terra-cotta warriors but home to numerous other antiquities, reached 400,000 people by A.D. 637. Its 8.5 million people make it as important a city now as then.

8. **Constantinople** (now Istanbul, Turkey): Emperor Constantine the Great made this city, first colonized by Greeks in the 1200s B.C, his eastern imperial Roman capital with 300,000 people. As Byzantium, it bobbed and wove through the tides of faith and conquest. Today, it is Turkey's largest city with more than 15.5 million people.

9. **Baghdad** (Iraq): Founded around A.D. 762, this center of Islamic culture and faith was perhaps the first city to house more than 1,000,000 people. Today, more than 7.5 million people call Baghdad home.

10. **Tenochtitlán** (now Mexico City, Mexico): Founded in A.D. 1325, this island-built Aztec capital had more than 200,000 inhabitants within a century. Most of the surrounding lake has been drained over the years. Today, more than 22 million souls call Mexico City home.

11. **Carthage** (now the ruins of Carthage, Tunisia): Phoenician seafarers from the Levant founded this great trade city in 814 B.C. Before the Romans obliterated it in 146 B.C., its population may have reached 700,000. Today, it sits in empty silence ten miles from modern Tunis—population 2.4 million.

22 Facts About Petra

1. The ancient city of Petra is a World Heritage Site and a Jordanian national treasure.

2. Located within the Kingdom of Jordan, Petra is some 80 miles south of Amman in the Naqab (Negev) Desert, about 15 miles east of the Israeli border.

3. Petra was a key link in the trade chain connecting Egypt, Babylon, Arabia, and the Mediterranean.

4. In 600 B.C., the narrow red sandstone canyon of Petra housed a settlement of Edomites: seminomadic Semites said to descend from the biblical Esau.

5. With the rise of the incense trade, Arab traders began pitching tents at what would become Petra. We know them as the Nabataeans.

6. The Nabataeans showed up speaking early Arabic in a region where Aramaic was the business-speak.

7. The newcomers thus first wrote their Arabic in a variant of the Aramaic script. But Petra's trade focus meant a need to adopt Aramaic as well, so Nabataeans did.

8. By the end (about 250 years before the rise of Islam), Nabataean "Arabaic" had evolved into classical (Koranic) Arabic.

9. The Nabataeans of Petra weren't expansionists, but defended their homeland with shrewd diplomacy and obstinate vigor. Despite great wealth, they had few slaves.

10. Petra's Nabataeans showed a pronounced democratic streak, despite monarchical government. Empires rose and fell around them; business was business.

11. The core commodity was incense from Arabia, but many raw materials and luxuries of antiquity also passed through Petra—notably bitumen (natural asphalt), useful in waterproofing and possibly in embalming.

12. Nabataean women held a respected position in society, including property and inheritance rights.

13. While no major ancient Near Eastern culture was truly egalitarian, the women of Petra participated in its luxuriant prosperity.

14. Most Nabataeans were pagan, worshipping benevolent fertility and sun deities.

15. Jews were welcome at Petra, as were Christians in its later days.

16. In 70 B.C., Petra was home to about 20,000 people.

17. Ornate homes and public buildings rivaling Athenian and Roman artistry were carved into the high red sandstone walls of the canyon.

18. Petra's last king, Rabbel II, willed his realm to Rome.

19. When he died in A.D. 106, Nabataea became the Roman province of Arabia Petraea. Again the Nabataeans adjusted—and kept up the trade.

20. In the 2nd and 3rd centuries A.D., the caravans began using Palmyra (in modern Syria) as an alternate route, starting a slow decline at Petra.

21. An earthquake in 363 delivered the knockout punch: damage to the intricate water system sustaining the city.

22. By about A.D. 400, Petra was an Arabian ghost town.

45 Cool Facts About Celts

1. Ancient Celts believed the head was the seat of the soul.

2. The Celts were fierce warriors from central Europe.

3. By 200 B.C. their civilization stretched across much of northern and western Europe.

4. The Celts did not have a single empire like the Romans, but instead lived in separate tribes with similar languages, religion, and customs.

5. Celts spoke related languages that survive today as Scottish Gaelic, Welsh, Irish, and Breton.

6. The Celts belonged to tribes ruled by kings, queens, and chieftains.

7. Classical Celtic culture emerged in central Europe around modern Austria, Bavaria, and Switzerland.

8. The earliest major Celtic settlement, dating from 1200 B.C., was found in Hallstatt, Upper Austria.

9. Celtic armies attacked Rome in 387 B.C. and invaded Greece in 279 B.C.

10. Some of these invaders then crossed into Galatia (modern-day Turkey) and built settlements there.

11. The Celts were expert metalworkers in iron, bronze, silver, and gold.

12. Iron was used for tools and weapons.

13. The Celts made metal rings called torcs that they wore around their necks.

14. The bravest Celtic warriors went into battle wearing nothing except a torc.

15. The Celts invented chainmail (around 300 B.C.) and helmets later used by Roman legionaries.

16. The ancient Celts used an early form of hair gel made from vegetable oil and resin from trees.

17. The ancient Celts traded goods with people in China.

18. The Celts believed in many gods and goddesses.

19. They sacrificed valuable objects and animals—and sometimes people—to the gods.

20. Ancient Celts threw precious objects into rivers and lakes, which they believed to be entrances to the world of the gods.

21. Celtic women had more rights than Greek or Roman women.

22. In A.D. 60, Queen Boudicca and her Iceni tribe rose in revolt against the Romans.

23. The walls of ancient Celtic homes were made of woven willow branches smeared with mud and animal dung.

24. The religious leaders of the Celts in Britain, Ireland, and Gaul (France) were called druids.

25. As the priestly class of Celtic society, the druids served as the Celts' spiritual leaders.

26. Druids were repositories of knowledge about the world and the universe, as well as authorities on Celtic history, law, religion, and culture. In short, they were the preservers of the Celtic way of life.

27. To become a druid, one had to endure as many as 20 years of oral education and memorization.

28. Celtic women could become druids.

29. They could also buy or inherit property, assume leadership, wage war, and divorce men.

30. In terms of power, the druids took a backseat to no one.

31. Even the Celtic chieftains, well-versed in power politics, recognized the overarching authority of the druids.

32. Celtic society had well-defined power and social structures and territories and property rights.

33. The druids were deemed the ultimate arbiters in all matters relating to such.

34. If there was a legal or financial dispute between two parties, it was unequivocally settled in special druid-presided courts.

35. Armed conflicts were immediately ended by druid rulings. Their word was final.

36. There were two forces to which even the druids had to succumb—the Romans and Christianity.

37. With the Roman invasion of Britain in A.D. 43, Emperor Claudius decreed that druidism throughout the Roman Empire was to be outlawed.

38. The Romans destroyed the last vestiges of official druidism in Britain with the annihilation of the druid stronghold of Anglesey in A.D. 61.

39. Surviving druids fled to unconquered Ireland and Scotland, only to become completely marginalized by the influence of Christianity within a few centuries.

40. In Scotland, the Celts made small artificial islands called crannogs in lakes and rivers, and built houses on them.

41. Most Celts lived in round houses with one central fire for heating, boiling water, and cooking.

42. Celtic warriors fought with iron swords and daggers, on foot or horseback.

43. The Romans and Greeks wrote about the Celts as swaggering barbarians who indulged in head-hunting and human sacrifices.

44. The Celts held large banquets to celebrate victories in battle. These lasted several days.

45. No one knows what the Celts called themselves. The term "Celt" is believed to come from Greek *Keltoi* or Latin *Celtae*.

18 Facts About Picts

1. The Picts inhabited ancient Scotland before the Scots.

2. Some scholars believe that the Picts were descendants of the Caledonians or other Iron Age tribes who invaded Britain.

3. No one knows what the Picts called themselves; the origin of their name comes from other sources and probably derives from the Pictish custom of tattooing or painting their bodies.

4. The Irish called them *Cruithni*, meaning "the people of the designs."

5. The Romans called them *Picti*, Latin for "painted people."

6. However, the Romans probably used the term for all the untamed peoples living north of Hadrian's Wall.

7. The Picts themselves left no written records.

8. All descriptions of their history and culture come from second-hand accounts.

9. The earliest of these is a Roman account from A.D. 297 stating that the Picti and the Hiberni (Irish) were already well-established enemies of the Britons to the south.

10. Before the arrival of the Romans, the Picts spent most of their time fighting amongst themselves.

11. The threat posed by the Roman conquest of Britain forced the squabbling Pict kingdoms to come together and eventually evolve into the nation-state of Pictland.

12. The united Picts were strong enough not only to resist conquest by the Romans, but also to launch periodic raids on Roman-occupied Britain.

13. Having defied the Romans, the Picts later succumbed to a more benevolent invasion launched by Irish Christian missionaries.

14. Arriving in Pictland in the late 6th century, they succeeded in converting the polytheistic Pict elite within two decades.

15. Much of the written history of the Picts comes from the Irish Christian annals.

16. If not for the writings of the Romans and the Irish missionaries, we might not have knowledge of the Picts today.

17. Despite the existence of an established Pict state, Pictland disappeared with the changing of its name to the Kingdom of Alba in A.D. 843, a move signifying the rise of the Gaels as the dominant people in Scotland.

18. By the 11th century, virtually all vestiges of the Picts had vanished.

33 Knights Templar Facts

1. The Crusades, Christendom's quest to recover and hold the Holy Land, saw the rise of several influential military orders, including the Knights Templar.

2. On July 15, 1099, the First Crusade stormed Jerusalem and slaughtered everyone in sight—Jews, Muslims, and Christians.

3. This unleashed a wave of pilgrimage, as European Christians flocked to now-accessible Palestine and its holy sites.

4. Though Jerusalem's loss was a blow to Islam, it was a bonanza for the region's thieves, as it brought a steady stream of naive pilgrims to rob.

5. French knight Hugues de Payen, with eight chivalrous comrades, swore to guard the travelers.

6. In 1119, they gathered at the Church of the Holy Sepulchre and pledged their lives to poverty, chastity, and obedience before King Baldwin II of Jerusalem.

7. The Order of Poor Knights of the Temple of Solomon took up headquarters in said Temple.

8. The Templars did their work well, and in 1127 King Baldwin sent a Templar embassy to Europe to secure a marriage that would ensure the royal succession in Jerusalem.

9. Not only did they succeed, influential nobles showered the Order with money and real estate, the foundation of its future wealth.

10. With this growth came a formal code of rules.

11. Templars could not desert the battlefield or leave a castle by stealth.

12. They had to wear white habits—except for sergeants and squires, who could wear black.

13. They had to tonsure (shave) their crowns and wear beards.

14. They were required to dine in communal silence, broken only by Scriptural readings.

15. Templars had to be chaste, except for married men joining with their wives' consent.

16. With offices in Europe to manage the Order's growing assets, the Templars returned to Palestine to join in the Kingdom's ongoing defense.

17. In 1139, Pope Innocent II decreed the Order answerable only to the Holy See.

18. Now the Order was entitled to accept tithes!

19. By the mid-1100s, the Templars had become a church within a church, a nation within a nation, and a major banking concern.

20. Templar keeps were well-defended depositories, and the Order became financiers to the crowned heads of Europe—even to the Papacy.

21. Their reputation for meticulous bookkeeping and secure transactions underpinned Europe's financial markets, even as their soldiers kept fighting for the faith in the Holy Land.

22. In 1187, Saladin the Kurd retook Jerusalem, martyring 230 captured Templars.

23. Factional fighting between Christians sped the collapse.

24. In 1291, the last Crusader outpost fell to the Mamelukes of Egypt. The Templars' troubles had just begun.

25. King Philip IV of France owed the Order a lot of money, and in 1307, Philip ordered all Templars arrested.

26. They stood accused of devil worship, sodomy, and greed.

27. Hideous torture produced piles of confessions.

28. The Order was looted and officially dissolved.

29. In March 1314, Jacques de Molay, the last Grand Master of the Knights Templar, was burned at the stake.

30. Many Templar assets passed to the Knights Hospitallers.

31. The Order survived in Portugal as the Order of Christ, where it exists to this day in form similar to British knightly orders.

32. A Templar fleet escaped from La Rochelle and vanished; it may have reached Scotland.

33. Swiss folktales suggest that some Templars took their loot and expertise to Switzerland, possibly laying the groundwork for what would one day become the Swiss banking industry.

23 Facts About the Maori

1. The Maori are a group of people who settled on the islands of New Zealand between A.D. 800 and 1200.

2. They had traveled across the South Pacific from the islands of Polynesia.

3. Maori tradition says they came from an island called Hawaiki, the mother island of the east Polynesians.

4. The Maori called the islands of New Zealand Aotearoa, which means "Land of the Long White Cloud."

5. Early Maori settlers of New Zealand hunted the moa, large flightless birds, to extinction.

6. Maori built canoes (*Waka*), which ranged in size from small one-man boats to large double-hulled war canoes called *Waka taua*, which were as long as 130 feet.

7. Traditional Maori clothing included skirts made from flax, a common plant found on New Zealand, and elaborate cloaks for warmth and to indicate status.

8. Te Reo Maori is the language of the Maori.

9. Today it is an official language of New Zealand.

10. Maori worship more than 70 gods, each strongly linked with nature.

11. According to legend, all Maori gods descended from Papatuanuku (Earth Mother) and Ranginui (Sky Father).

12. The Maori creation story describes the world being formed by the violent separation of Ranginui and Papatuanuku by their children.

13. Many Maori carvings and artworks graphically depict this struggle.

14. Maori society was divided into a number of large tribes known as *iwis*.

15. Each *iwi* consisted of a number of smaller tribes called *hapus*, with each led by an *ariki*, or chief.

16. Each *hapu* was further divided into *whanaus*, or families.

17. The families came together as a *hapu* for rituals and activities in a common meetinghouse known as a *wharenui*.

18. Elaborately carved meetinghouses were the pride of the tribe.

19. Tiki motifs often decorated meetinghouses.

20. A feast centering around a pit in the ground was a cultural tradition of the Maori.

21. Food would be put in ovens called *hangi*.

22. Heated stones would cook the meat and vegetables.

23. Maori held eggs in high esteem and would place them in the hands of the dead before burial.

Chapter 3

ANIMALS

32 Freaky Facts About Animal Mating

1. The funnel-web spider knocks his mate unconscious with pheromones before mating.

2. The male *Argyrodes zonatus* spider secretes a drug that intoxicates the female, which is good because otherwise she would devour him.

3. Harlequin bass and hamlet fish take turns being male and female, including releasing sperm and eggs during mating.

4. Male North American fireflies flash their light every 5.8 seconds while females flash every 2 seconds, so there isn't any confusion.

5. An albatross can spend weeks courting, and their relationships can last for decades.

6. However, the actual act of mating lasts less than a minute.

7. Fruit flies perform an elaborate seven-step dance routine before mating.

8. No copulation takes place unless each part is completed perfectly.

9. Male mites mate with their sisters before they are born.

10. After birth, the females rush off in search of food and their brothers are left to die.

11. Mayflies live for one day, during which they do nothing but mate.

12. The rattlesnake has two penises, while that of the echidnas has four heads and a pig's is shaped like a corkscrew.

13. The male swamp antechinus, a mouse-like marsupial in Australia, has sex until he dies, often from starvation.

14. Sometimes he's simply too weak after mating to escape predators.

15. It's no fun being a male honeybee. Those "lucky" enough to engage in a mating flight with a virgin queen usually die after their genitalia snap off inside her.

16. Flatworms are hermaphrodites, having both male and female sex organs.

17. During mating, two worms will "fence" with their penises until one is pierced and impregnated.

18. During mating season, male frigate birds inflate their throat sacs while engaging in a wild dance.

19. The females usually hook up with the males possessing the largest, brightest sacs.

20. When it comes time to mate, male Galapagos giant tortoises rise on their legs and extend their necks; the male with the longest neck gets the girl.

21. When a female red-sided garter snake awakens from hibernation, she releases a scent that attracts every male in the area. The result: a huge, writhing "mating ball."

22. Male giraffes won't mate unless they know a female is in estrus.

23. To find out, they nudge a prospective mate's rump until she urinates—then taste her urine.

24. Percula clownfish live in families consisting of a mating male and female and several nonbreeding males.

25. If the female dies, the mating male becomes the female, and one of the nonbreeding males gets promoted to hubby.

26. Male dolphins have such strong libidos that they've been known to mate with inanimate objects and even with other sea creatures, such as turtles.

27. When the male snowy owl wishes to arouse a female, he dances while swinging a dead lemming from his beak.

28. Amorous great gray slug couples hang from a rope of their slime as they twist around each other in the throes of slug passion.

29. In a Mediterranean species of cardinal fish, the male takes part in mouthbrooding—holding the fertilized eggs in his mouth until they're ready to hatch.

30. All shrimp form harems, usually consisting of one male and up to 10 females.

31. When the male leader dies, he may be replaced by a young female that can change her gender to take his place.

32. After the female emperor penguin lays an egg, the male protects it in his brood pouch—a roll of skin and feathers between his legs that drops over the egg—until it hatches.

29 Elephant Facts

1. The Asian elephant is endangered, while the African elephant is vulnerable.

2. A female elephant is pregnant for 22 months—almost two years. Yikes.

3. Depending on the weather, an Asian elephant can guzzle 30 to 50 gallons of water every day.

4. The largest known specimen of the African savanna elephant is on display at the Smithsonian's National Museum of Natural History in Washington, D.C.

5. It stood 13 feet tall and weighed 22,000 pounds when it was alive.

6. Elephants never forget. They can recall distant watering holes, other elephants, and humans they've encountered, even after many years.

7. The average male Asian elephant weighs between 10,000 and 12,000 pounds.

8. African savanna elephants are currently most common in Kenya, Tanzania, Botswana, Zimbabwe, Namibia, and South Africa.

9. Elephants supplement the sodium in their food by visiting nearby mineral licks.

10. The closest relatives to the elephant are the hyraxes (small chunky mammals that resemble fat gophers), dugongs and manatees, and aardvarks.

11. Several physical characteristics distinguish Asian and African elephants. Asian elephants are generally smaller, with shorter tusks. They also have two domed bulges on their foreheads, rounded backs, less wrinkly skin, and their trunks have a finger-like projection at the tip.

12. African forest elephants live in central and western Africa, with the largest populations found in Gabon and the Republic of the Congo.

13. African elephants speak their own special language.

14. Communication takes place using rumbles, moans, and growls. These low-frequency sounds can travel a mile or more.

15. An elephant's trunk comprises 150,000 different muscle fibers.

16. Ever wonder why an elephant has a trunk? It's because an elephant's neck is so short that it wouldn't be able to reach the ground. The trunk allows the elephant to eat from the ground as well as the treetops.

17. Both female and male African elephants have tusks, but only male Asian elephants have tusks.

18. An elephant's trunk weighs about 400 pounds.

19. Between 1979 and 1989, Africa's elephant population went from 1.3 million to 750,000, as a result of ivory poaching.

20. About 415,000 African elephants remained as of 2016.

21. Elephants greatly enjoy baths, but in the meantime the red-billed oxpecker (a relative to the starling) picks ticks and other parasites off the elephant's skin.

22. Those big, floppy ears aren't just for decoration—they help the huge animals cool off.

23. A group of elephants is called a herd.

24. When two herds join forces (which happens during migration) it's called a clan.

25. Elephants were first seen in Europe in 280 B.C., when an army of 25,000 men and 20 elephants crossed from North Africa to Italy.

26. Adult elephants can't jump.

27. Elephants use mud as sunscreen.

28. Under the right conditions, an elephant can smell water from approximately three miles away.

29. Elephants sometimes "hug" by wrapping their trunks together as a way of greeting one another.

15 Odd Ostrich Facts

1. Contrary to popular belief, ostriches don't bury their heads in the sand.

2. Ostriches can't fly, but they can run at speeds of 45 miles per hour for up to 30 minutes.

3. The African black ostrich (*Struthio camelus domesticus*) is farmed for meat, leather, and feathers.

4. Ostriches have the best feed-to-weight ratio gain of any farmed land animal in the world and produce the strongest commercially available leather.

5. The ostrich's eyes are about the size of billiard balls.

6. An ostrich's brain is smaller than either of its eyeballs. This may explain why it tends to run in circles.

7. The ostrich's intestines are 46 feet long—about twice as long as those of a human.

8. Ostriches in captivity have been known to swallow coins, bicycle valves, alarm clocks, and even small bottles.

9. The ostrich is the largest living bird, standing six to ten feet tall and weighing as much as 340 pounds.

10. This great bird has only two toes on each foot; all other birds have three or four.

11. Ostriches kick forward, not backward, because that's the direction in which their knees bend.

12. Their powerful, long legs can deliver lethal kicks.

13. Although ostrich eggs are the largest of any bird, they are the smallest eggs relative to the size of the adult bird.

14. A three-pound egg is only about 1 percent as heavy as the ostrich hen.

15. Ostriches never need to drink water—some of it they make internally, and the rest is derived from their food.

46 Rodent Facts

1. Rodents are the largest group of mammals, constituting nearly half of the class Mammalia's species.

2. The world's 4,000 known rodent species are divided into three suborders: squirrel-like rodents (Sciuromorpha),

mouse-like rodents (Myomorpha), and porcupine-like rodents (Hystricomorpha).

3. Rodents include not only "true" rats and mice, but also such diverse groups such as beavers, marmots, pocket gophers, and chinchillas.

4. Chinchillas have the thickest fur of all land animals with more than 50 hairs growing from a single follicle.

5. Like other rodents, a chinchilla's teeth never stop growing.

6. Chinchillas are native to the Andes Mountains in Chile; their thick fur allows them to survive harsh winds and plunging temperatures at elevations of 12,000 feet.

7. Some people keep chinchillas as pets. Only a few thousand survive in the wild.

8. Mice can get by with almost no water; they get most of the moisture they need from their food.

9. Mice become sexually mature at six to ten weeks and can breed year-round.

10. Female mice average six to ten litters annually.

11. If a pair of mice started breeding on January 1, they could have as many as 31,000 descendants by the end of the year.

12. Measuring over two feet tall, four feet long, and weighing an average of 100 pounds, the South American capybara is the largest rodent in the world.

13. Capybaras love the water and can remain submerged for up to five minutes.

14. Rats can jump three feet straight up and four feet outward from a standing position.

15. Rodents spread an array of diseases, including bubonic plague, leptospirosis, tularemia, salmonellosis, murine typhus, and hantavirus.

16. The bubonic plague that killed millions throughout Europe in the mid-1300s was predominantly caused by fleas that rats carried.

17. Thousands of Americans are bitten by rats each year.

18. Rats constantly gnaw anything softer than their teeth, including bricks, wood, and aluminum sheeting.

19. The average lifespan of a rat is less than three years, but one pair can produce 2,000 offspring in a year.

20. Rats use their tails to regulate their temperature, communicate, and balance.

21. Some rats can swim as far as half a mile in open water, dive through water-plumbing traps, travel in sewer lines against strong currents, and stay underwater for as long as three minutes.

22. A single rat can produce 25,000 droppings in a year.

23. Rats can enter a building through a hole just half an inch.

24. A rat can fall 50 feet without injury.

25. Rats can jump 36 inches vertically and 48 inches horizontally.

26. Rats are intelligent and have excellent memories. Once rats learn a route, they never forget it.

27. Rats are color blind and cannot vomit or burp.

28. Beavers are the largest rodents in the world after the South American capybara.

29. Transparent eyelids allow beavers to see while swimming underwater.

30. Beavers can stay submerged in water for up to 15 minutes.

31. Using their webbed feet for speed and their flat tail as a rudder, they can swim as fast as five miles per hour.

32. A single beaver can fell an aspen tree with a six-inch diameter in about 20 minutes.

33. It then gnaws the tree into logs of a more manageable size and drags these logs back to the river.

34. Beavers live in colonies of six to eight and build dams and the lodges they live in.

35. Porcupines look a little like hedgehogs, but they are not related. In fact, hedgehogs aren't rodents at all.

36. The average porcupine has 30,000 quills. And each is a sharp reminder that if you poke at a porcupine, it will poke you back!

37. Baby porcupines are born with soft quills that don't harden until a few days after birth— a fact for which mother porcupines are supremely grateful.

38. Porcupines can float.

39. The Baluchistan pygmy jerboa is the world's smallest rodent at only one-and-a-half inches long.

40. Pocket gophers, named for their large, fur-lined cheek pockets, can close their lips behind their protruding, chisel-like front teeth. This allows them to excavate soil without ingesting it.

41. Voles have blunt noses, small furry ears, dense brown fur, and a tail with no fur.

42. Voles typically live about three to six months.

43. Marmots are giant land squirrels.

44. Flying squirrels glide through the air using parachute-like wings.

45. A squirrel can smell a nut buried under a foot of snow.

46. The naked mole-rat, a burrowing rodent native to East Africa, is almost completely hairless with wrinkly skin.

23 Fun Fish Facts

1. There are more than 30,000 different species of fish.

2. Three major types of fish include jawless (e.g. lamprey eels), cartilaginous (e.g. sharks), and bony (e.g. blue marlin).

3. While fossils show that the earliest fish were jawless, the only remaining survivors in the jawless fish group are hagfish and lampreys.

4. Hagfish feed on dead fish at the ocean bottom, using their tongues to rasp at food with a pair of brushes covered in hornlike teeth.

5. Hagfish are sometimes called slime hags because of the large amounts of mucus they produce.

6. One hagfish can fill a two-gallon bucket with slime in a matter of minutes.

7. All lampreys start life as freshwater larvae, filtering particles from the bottom of riverbeds before developing teeth as adults.

8. The heaviest bony fish is the ocean sunfish, which can weigh a whopping 5,000 pounds.

9. The ocean sunfish produces about as many eggs at one time as there are people in the United States.

10. Of all fish, seahorses are the slowest.

11. Some seahorses swim less than five feet per hour.

12. Australian leafy sea dragons mask themselves as seaweed.

13. The green creatures have long, tattered ribbons of skin resembling seaweed fronds growing from their bodies.

14. Scientists can figure out how old a fish is by counting growth rings on its scales or its otoliths (ear bones).

15. The orange roughy lives more than 100 years.

16. The Mariana snailfish (*Pseudoliparis swirei*) thrives at depths up to 26,200 feet below the surface.

17. These deep-ocean dwellers were discovered along the Mariana Trench.

18. Some species can swim backwards, but usually don't. Those that can are mostly eels.

19. An electric eel can produce 600 volts of electricity in a single jolt.

20. Whale sharks are the longest fish in the sea, with some growing over 40 feet long.

21. At just nine millimeters long, the dwarf goby fish is the world's smallest fish.

22. Sailfish can swim as fast as 68 miles per hour.

23. Goldfish only have a memory of three seconds.

22 Facts About Bats

1. Bats are the only mammals that can truly fly.

2. A quarter of all mammals are bats.

3. There are over 1,400 species of bats worldwide.

4. A single brown bat can eat up to 1,200 mosquito-size insects in one hour.

5. Because bats eat so many insects, which have exoskeletons made of a shiny material called chitin, some bat poop sparkles.

6. Bats are nocturnal, mostly because it's easier to hunt bugs and stay out of the way of predators when it's dark.

7. Bats use echolocation to navigate in the dark: They send out beeps and listen for variations in the echoes that bounce back at them.

8. Inside those drafty caves they like so much, bats keep warm by folding their wings around them, trapping air against their bodies for instant insulation.

9. Vampire bats don't suck blood. They puncture their prey's skin with sharp incisors, then lap up the flowing blood.

10. Vampire bat saliva contains an anticoagulant that prevents blood from clotting too quickly.

11. There are three species of vampire bats, which only drink blood: the common vampire bat, the hairy-legged vampire bat, and the white-winged vampire bat.

12. Bats have only one pup a year: Most smallish mammals have way more offspring.

13. Over 300 species of fruit—including bananas, mangoes, and avocadoes—rely on bats for pollination.

14. The rare suckerfooted bat of Madagascar has small suction cups on its hands, allowing it to cling to smooth surfaces as it glides through forests.

15. The *Anoura fistulata* nectar bat has a longer tongue relative to its body length than any other mammal.

16. When the bat is not using its tongue to reach inside flowers and get pollen, the appendage curls up in the bat's rib cage.

17. Bats clean themselves and each other meticulously by licking and scratching for hours.

18. The world's smallest mammal is Kitti's hog-nosed bat, also called the bumblebee bat. It weighs only as much as a dime.

19. Flying foxes are among the world's largest bats with wingspans of up to six feet.

20. Millions of bats have died in recent years from white-nose syndrome, named for the white fungus on bats' muzzle and wings.

21. Some species of bats can live more than 30 years.

22. Nearly 1.5 million bats living in North America's largest urban bat colony call the Congress Avenue Bridge in Austin, Texas, home.

45 Marvelous Mammal Facts

1. Mammals are warm-blooded animals that breathe air.

2. Zebras are black with white stripes (not white with black stripes) and have black skin.

3. A polar bear's fur is actually transparent rather than white; it merely appears white due to the way it reflects light.

4. Howler monkeys are the loudest land animals; their calls can be heard up to three miles away.

5. The Japanese macaque is that rare monkey that likes cold weather. When temperatures in its mountain habitat drop below freezing, the macaque lounges in natural hot springs.

6. It's not water in a camel's hump. The fat stored in a camel's hump allows the animal to trek across the desert for up to a month without food.

7. Pigs, light-colored horses, and walruses can get sunburned.

8. Measuring up to 110 feet long and weighing up to 419,000 pounds, the blue whale is the largest mammal (and animal) on Earth.

9. Most elephants weigh less than the tongue of a blue whale.

10. Just as humans favor their right or left hand, elephants favor their right or left tusk.

11. The echidna is a spiny anteater that can grow to three feet long.

12. The echidna and the platypus are the only egg-laying mammals.

13. The platypus is also one of the few venomous mammals.

14. The male platypus has a spur on his hind feet that can deliver venom.

15. The pangolin is the only mammal with scales.

16. It has no teeth, and uses its powerful claws to tear open termite and ant mounds.

17. The African honey badger's tough hide can resist penetration and most poisons, a great help when the badger's dinner includes puff adders or beehive honey.

18. Hedgehogs have 3,000 to 5,000 quills on their backs.

19. A mature ewe yields seven to ten pounds of shorn wool per year.

20. The sleepiest mammals are armadillos, sloths, and opossums. They spend 80 percent of their lives dozing.

21. The nine-banded armadillo is the only mammal known to always give birth to four identical young.

22. The tallest mammals are giraffes, towering up to 20 feet tall.

23. A giraffe can clean its ears with its 21-inch-long tongue.

24. The dark purple color of a giraffe's tongue protects it from getting sunburned.

25. Both male and female caribou grow antlers—the only deer species to do so.

26. The red deer on the island of Rhum (in Scotland) kill seabird chicks and gnaw the bones to get nutrients otherwise unavailable on the isle.

27. Rhinoceroses are odd-toed hoofed mammals, with three toes on each foot.

28. Horses and zebras have only one toe on each foot.

29. Even-toed hoofed mammals include pigs, deer, cattle, antelopes, giraffes, camels, and hippopotamuses.

30. The hippopotamus can open its mouth up to 3.3 feet wide—wider than any other land animal.

31. Hippos can run at speeds of up to 20 miles per hour.

32. The rabbit-size mouse deer of Asia has long upper incisors that make it look like a vampire. But these timid creatures quickly flee when they encounter people.

33. Koalas aren't bears. Koalas are marsupials, which means they raise their young in special pouches.

34. Eucalyptus leaves are poisonous to most animals, but koalas have special bacteria in their stomachs that break down dangerous oils in the leaves.

35. A koala gets almost all the liquid it needs from licking dew off tree leaves.

36. Some kangaroos can jump five times their body length in one jump.

37. Tiny shrews, sometimes only a few inches long, can kill prey twice their size.

38. They are able to do this partly because their saliva contains a paralyzing substance similar to cobra venom.

39. Cows and other ruminants have four digestive chambers.

40. An opossum will empty its anal glands when "playing dead" to help it smell like a rotting corpse.

41. Wolverines may look like small bears, but they're actually the largest species of weasel.

42. The Tasmanian devil can give birth to as many as 30 live young at one time.

43. Because a mama Tasmanian devil has only four teats, her babies compete fiercely for milk. Eventually, Mom ends up eating many of her kids.

44. The most widespread meat-eating mammals in the world are red foxes. Their natural range includes much of the Northern Hemisphere.

45. Mammals have better hearing than non-mammals thanks to special ear bones.

28 Amazing Ant Facts

1. Ants outnumber humans a million to one. Their combined weight outweighs the combined weight of all the humans.

2. Ants are found on every continent except Antarctica.

3. Ants can lift 10 to 50 times their body weight.

4. Ants don't have ears. They "hear" by feeling vibrations in the ground through their feet.

5. There are three kinds of ants in a colony: the queen, the female workers, and the males.

6. Depending on the species, a colony may have one queen or multiple queens.

7. Male ants' only job is to mate with the queen. Males die soon afterwards.

8. Only a queen ant can lay eggs. During a queen's 10- to 30-year reign, she may lay millions of eggs.

9. A queen's fertilized eggs become females; unfertilized eggs become males.

10. Most females are born sterile, consigned to be workers.

11. Ants pass through four life stages: egg, larvae, pupae, and adult.

12. The tiny ant eggs are sticky, allowing them to bond together for ease of care.

13. Young workers care for the queen and larvae, then graduate to nest duties such as engineering, digging, and sanitation.

14. Finally, they advance to the dangerous positions of security and foraging.

15. When foraging, ants leave a pheromone trail to show where they've been.

16. Only four species engage in agriculture: humans, termites, bark beetles, and ants. But ants were the first.

17. Leafcutter ants carefully cultivate subterranean fungus gardens by spraying their crops with self-produced antibiotics to ward off disease, then fertilizing them with their protease-laced anal secretions.

18. Leafcutter colonies can contain as many as two million ants.

19. A leafcutter colony can strip a citrus tree of its leaves in less than a day.

20. Ants engage in livestock farming. They domesticate and raise aphids, which they milk for honeydew.

21. Honeydew provides important nourishment for ants, which are incapable of chewing or swallowing solids.

22. Though blind, nomadic South American army ants fearlessly attack reptiles, birds, small mammals, and other insects (which they kill but don't eat) in their path.

23. There are over 12,000 ant species in the world.

24. The barbarous *Polyergus rufescens* species, or slave-maker ants, raid neighboring nests to steal their young.

25. No males of the species *Mycocepurus smithii* have yet been found.

26. The queen ant reproduces asexually, so all offspring are clones of the queen.

27. Carpenter ants build their homes, called galleries, within wood. They prefer slightly moist wood.

28. Ants stretch when they wake up. Ants also appear to yawn in a very human manner before taking up the tasks of the day.

11 Facts About Opossums

1. The rat-tailed opossum is the only marsupial native to North America.

2. Contrary to popular belief, opossums do not sleep hanging from trees by their tails.

3. Opossums possess a whopping 50 teeth—more than any other land-dwelling North American mammal.

4. The lowly opossum has opposable, thumblike digits on all four paws and a tail that can grasp food and tree branches.

5. A male opossum is called a jack.

6. A female opossum is called a jill.

7. After a gestation of just 12 to 13 days, females give birth to as many as 20 live young at a time.

8. Baby opossums, called joeys, are the size of jelly beans at birth.

9. After baby opossums leave the pouch (at between two and three months), the mother carries all of them on her back for the next month or so whenever the family leaves its den.

10. Opossums really do play dead to trick predators.

11. Opossums are immune to snake bites, bee stings, and other toxins.

23 Amphibian Facts

1. Amphibians live both on land and in water.

2. The word amphibian comes from a Greek term meaning "double life."

3. Amphibians include frogs and toads, salamanders and newts, and wormlike caecilians.

4. The largest caecilians can grow more than five feet in length.

5. Axolotls are rare amphibians that live in Central American lakes.

6. All amphibians have gills, some only as larvae and others for their entire lives.

7. The largest amphibian in the world is the Chinese giant salamander, which can grow up to six feet long.

8. The world's smallest amphibian is the *Paedophryne amauensis*, a frog from Papua New Guinea measuring just 0.3 inch in length.

9. Amphibians are found everywhere in the world except Antarctica.

10. Some salamanders can regrow entire limbs and regenerate parts of major organs.

11. At about four weeks old, tadpoles get a bunch of very tiny teeth, which help them turn their food into mushy, oxygenated particles.

12. Frogs swallow their food whole.

13. The size of what they can eat is determined by the size of their mouth and their stomach.

14. While swallowing, a frog's eyeballs retreat into its head, applying pressure that helps push food down its throat.

15. Some frogs glow after eating fireflies.

16. When frogs aren't near water, they will often secrete mucus to keep their skin moist.

17. Frogs typically eat their old skin once it's been shed.

18. The Goliath frog of West Africa is the largest frog in the world. When fully stretched out, it is often more than 2.5 feet long!

19. A small, blind cave salamander called the olm (also called "the human fish" for its humanlike skin tone) is the world's longest-lived amphibian. It can live up to 100 years.

20. A frog's ear is connected to its lungs.

21. When a frog's eardrum vibrates, its lungs do, too. This special pressure system keeps frogs from hurting themselves when they blast loud mating calls.

22. Wood frogs can continuously freeze and thaw throughout the winter in order to hibernate under surface leaves.

23. Just one golden poison frog has enough toxin to kill ten people.

23 Surprising Facts About Spiders

1. Spiders are arachnids, not insects.

2. Unlike insects, spiders cannot fly.

3. After they hatch, spiderlings drift away on silk string.

4. This first flight is called "ballooning."

5. Some young spiderlings have been seen ballooning at altitudes of 10,000 feet.

6. Studies have shown that jumping spiders can solve simple 3-D puzzles; they also learn the behavior patterns of other spiders in order to capture them.

7. Male spiders are unique among all animals in having a secondary copulatory organ.

8. While most spiders live for a year or two, certain tarantulas can live for 20 years or longer.

9. Male spiders are almost always smaller than the females and are often much more colorful.

10. It is estimated that up to 1 million spiders live in one acre of land—in the tropics, that the number might be closer to 3 million.

11. Some spiders live underwater all of their lives. They surface to collect a bubble of air, which acts as an underwater lung.

12. The fisher, or raft, spider can walk on water.

13. When it detects prey (insects or tiny fish) under the surface, it can quickly dive to capture its dinner.

14. Spiders eat more insects than birds and bats eat combined.

15. Some species of spiders have evolved to mimic ants in appearance and pheromones, which they use to prey on the ants.

16. The body of a typical spider has approximately 600 silk glands.

17. Spider silk is more flexible than nylon, and in some cases stronger than steel and Kevlar.

18. All spiders secrete silk from their abdomens. Tarantulas can also shoot silk from their feet.

19. Most spiders are nearsighted. To compensate, they rely on their body hair to feel their way around and to detect nearby animals.

20. While most spiders have eight eyes, some have fewer, though always an even number.

21. The world's most venomous spider is the male Sydney funnel-web spider *Atrax robustus*.

22. The female is far less dangerous, being four-to-six times less effective than the male.

23. The black widow is North America's most venomous spider.

50 Animals by the Bunch

1. A congregation of alligators

2. A shrewdness of apes

3. A cete of badgers

4. A battery of barracudas

5. A kaleidoscope of butterflies

6. A wake of buzzards

7. A gang or obstinacy of buffalo

8. A quiver of cobras

9. A bask of crocodiles

10. A convocation of eagles

11. A parade of elephants

12. A cast of falcons

13. A business of ferrets

14. A charm of finches

15. A flamboyance of flamingoes

16. A leash or skulk of foxes

17. A gaggle of geese

18. A tower of giraffes

19. A troubling of goldfish

20. A bloat of hippopotamuses

21. A cackle of hyenas

22. A mess of iguanas

23. A husk of jackrabbits

24. A shadow of jaguars

25. A smack of jellyfish

26. A mob or troop of kangaroos

27. A conspiracy of lemurs

28. A leap of leopards

29. A barrel of monkeys

30. A murder of magpies

31. A passel of opossums

32. A family or romp of otters

33. A parliament of owls

34. A pandemonium of parrots

35. A covey of partridges

36. An ostentation of peacocks

37. A prickle of porcupines

38. A bevy of quails

39. A gaze of raccoons

40. A mischief of rats

41. A rhumba of rattlesnakes

42. An unkindness of ravens

43. A shiver of sharks

44. A stench of skunks

45. A dray or scurry of squirrels

46. A fever of stingrays

47. An ambush or streak of tigers

48. A knot of toads

49. A wisdom of wombats

50. A zeal of zebras

43 Remarkable Reptile Facts

1. Reptiles are cold-blooded animals covered in scales or bony plates. They include lizards, turtles, snakes, and crocodiles.

2. Lizards are the most common reptiles, with nearly 6,000 species around the world.

3. Chameleons change color with the help of special skin cells called chromatophores.

4. Chameleons use their color-changing ability mainly to show their mood, adjust their temperature, or communicate.

5. Chameleons use each eye independently, so they can look in two directions at once.

6. Leaf chameleons of Madagascar are the world's smallest reptile, measuring just over an inch from nose to tail.

7. The Komodo dragon is the world's biggest lizard.

8. An average male weighs 175 to 200 pounds and measures 8.5 feet long.

9. The shingleback lizard of Australia has a thick, short, rounded tail that's shaped just like its head.

10. This confuses predators, giving the slow, sleepy reptile a better chance to escape.

11. Horned lizards, often called horny toads, can squirt blood from their eyeballs to attack predators.

12. This only happens in extreme cases, but they can shoot it up to three feet.

13. The longest reptile is the reticulated python, measuring over 32 feet.

14. The world's smallest snake is the Martinique thread snake, which measures just over four inches long when fully grown.

15. Some snakes have more than 300 pairs of ribs.

16. Snakes have a taste sensor on the roof of their mouth.

17. They use their tongue to pick up scents in the air, then deliver them back to the sensor inside their mouth.

18. Snakes have no eyelids.

19. Several species of snakes, including the spitting cobra, fake death by flipping over on their backs when threatened.

20. The female boomslang, a type of African tree snake, looks so much like a tree branch that birds—its main prey—will land right on it.

21. Boas and pythons kill by squeezing, or constricting, their prey.

22. These constrictors wrap themselves around their prey and use their powerful muscles to suffocate them.

23. Vipers inject their prey with venom through their hollow, needle-pointed fangs.

24. Venomous snakes kill more than 100,000 people each year.

25. Female pythons keep their eggs warm by coiling on top of them and shivering their bodies to generate heat.

26. Tortoises live on land, while turtles live mostly in water.

27. The marine leatherback turtle can weigh as much as 1,800 pounds with an 8-foot shell.

28. Sea turtles can hold their breath for several hours.

29. Sea turtles absorb a lot of salt from the sea water in which they live. They excrete excess salt from their eyes, so it often looks as though they are crying.

30. Giant tortoises can live more than 200 years.

31. The ribs and backbones of turtles and tortoises are connected to their shell.

32. Shells are made of solid plates called scutes, which are fused together. The domed top is called the carapace.

33. Alligators lack chromosomes for sex determination.

34. Hot temperatures during incubation produce males, while cooler temperatures produce females.

35. Crocodiles bury their eggs in riverside nests. The mother waits nearby for up to three months.

36. Crocodiles and alligators are surprisingly fast on land, but lack agility. If you're being chased by one, run in a zigzag line.

37. Unborn crocodiles make noises inside their eggs to alert the mother that it's time to hatch.

38. The mother then digs down to the nest and carries her babies to the water inside her mouth.

39. A mother crocodile can carry as many as 15 babies in her mouth at one time.

40. Female Nile crocodiles often gently roll the eggs in their mouth to crack the shell and help hatching babies emerge.

41. While crocodiles have large teeth and powerful jaws, they lift their young with incredible care.

42. As an alligator's teeth wear down, they are replaced with new ones. As a result, the average alligator can go through 2,000 to 3,000 teeth over its lifetime.

43. Earth's largest crocodilian is the saltwater crocodile, which can grow as long as 23 feet and weigh as much as 2,200 pounds.

50 Stupendous Shark Facts

1. Sharks have been around for over 400 million years.

2. Sharks have no tongues.

3. Their taste buds are in their teeth.

4. Sharks have different types of teeth. Mako sharks have very pointed teeth.

5. White sharks have triangular, serrated teeth.

6. A sandbar shark will have around 35,000 teeth over the course of its lifetime.

7. The tiny cookiecutter shark, rarely seen by human eyes, has big lips and a belly that glows a pale blue-green color to help camouflage it from prey.

8. Its name comes from the small, cookie-shape bite marks it leaves.

9. Sharks have no bones. Their skeletons are made of cartilage.

10. Hammerhead sharks have 360-degree vision.

11. Greater hammerheads are known to group in large schools of 100 or more off the Island of Cacos near Mexico.

12. Shark skin feels like sandpaper because it's covered in tiny, pointed structures called placoid scales.

13. Sharks are known for their "sixth sense."

14. They can sense electromagnetic fields of other creatures and objects and temperature shifts thanks to special sensors called the Ampullae of Lorenzini.

15. A shark can be put into a catatonic state called "tonic immobility" when it's flipped onto its back or when the Ampullae of Lorenzini are appropriately stimulated.

16. Several types of sharks have demonstrated an affinity for being touched or for being put into a state of tonic immobility.

17. Humans are much more dangerous to sharks than sharks are to humans.

18. People kill as many as 100 million sharks every year, often when the sharks are accidentally caught in fishing nets.

19. Many other sharks are caught only for their fins, which are used in shark-fin soup.

20. Tiger sharks are often called the "garbage cans of the sea" because they will eat nearly anything.

21. Contents of their stomachs have revealed tires, baseballs, and license plates.

22. Shortfin makos are the fastest sharks.

23. They have been clocked at 36 miles per hour and have been estimated to swim up to 60 miles per hour.

24. They need this extreme speed to chase down their favorite food—the lightning-fast yellowfin tuna.

25. Scientists have identified more than 400 species of sharks in the world.

26. Approximately 30 of those species are considered dangerous to humans.

27. Bull sharks, one of the most dangerous, aggressive shark species, have the highest testosterone levels of any animal in the world.

28. Bull sharks can survive in both saltwater and freshwater.

29. Whale sharks are the biggest fish in the ocean.

30. They can grow up to 40 feet and weigh more than 74,000 pounds.

31. Despite their large size, whale sharks are no danger to humans. They feed on mostly plankton and small fish.

32. A whale shark's spot pattern is unique as a fingerprint.

33. Basking sharks are the world's second largest fish, growing as long as 32 feet and weighing more than five tons.

34. The pygmy ribbontail catshark is the smallest shark in the world, with a maximum length of seven inches.

35. Most sharks are solitary hunters, but some are quite social.

36. Blacktip reef sharks frequently hunt in packs, helping one another grab fish and crabs out of the coral.

37. The great white shark can go nearly three months without food.

38. Great white attacks are usually caused by the animals' curiosity about an unfamiliar object.

39. Lacking hands, the curious creatures "feel out" the new object with their teeth, usually in a gentle bite.

40. A person is 1,000 times more likely to be bitten by a dog than by a shark, and dogs kill more people every year than sharks do.

41. If you happen to be attacked by a shark, try to gouge its eyes and gills, its most sensitive areas.

42. Sharks are opportunistic feeders and generally don't pursue prey that puts up a fight in which they could be injured.

43. Gansbaii, South Africa, touts itself as the "Great White Capital of the World."

44. Its shores host the greatest concentration of great white sharks in any ocean.

45. An extinct relative of the great white, *Carcharodon megalodon* grew to 50 feet long.

46. They died out around a million years ago.

47. Not all sharks live in the ocean. River sharks have been found in rivers in South Asia, New Guinea, and Australia.

48. The number of pups in a litter varies by species. A blue shark once gave birth to 135 pups in a single litter.

49. Female sharks have been known to use the sperm from multiple males when they reproduce—meaning that pups they give birth to at the same time may be just half-siblings.

50. The spiny dogfish shark can take two years to gestate before delivery—making it the longest gestation period of any vertebrate.

10 Mating Habits

1. **Prairie Voles:** The male of this rodent breed prefers to stay with the female he loses his virginity to and will even attack other females.

2. **Bald Eagles:** Like all raptors, including the golden eagle, hawk, and condor, bald eagles remain faithful until their mate dies.

3. **Penguins:** Penguins practice serial monogamy. This means they have several mates throughout their lives

but only one at a time. Penguins usually switch it up each mating season.

4. **Wolves:** Monogamy makes it easier for alphas to display their strength and superiority over the other male members of the wolf pack. In a grey wolf pack, the male and female alpha mate for life, cementing their position as pack leaders.

5. **Anglerfish:** Like a parasite, the male bites into the female, fusing his mouth to her skin. Their bloodstreams merge together, and the male hangs there, slowly degenerating until he becomes nothing more than a source of sperm for the female.

6. **Beavers:** Eurasian beavers team up for life as a way to increase their chances of survival. By pairing up, couples can split their workload.

7. **Black Vultures:** Gossip gets around—just ask a vulture. If one of them gets caught mating with a bird that's not its partner, it gets harassed not only by its "spouse" but also by all the other vultures in the area.

8. **Gibbon Apes:** These monkeys make a close-knit family unit, with mom, dad, and babies even traveling together as a group.

9. **Red-Backed Salamanders:** Males of these species are the jealous type, physically and sexually harassing their female mates if they suspect infidelity or even see them associating with another male. Yikes.

10. **Termites:** Termites can mate for life (that's about five years) and raise a family together—but they often leave each other within the first two hours of mating if something better comes along.

31 Baby Animal Facts

1. Just like humans, all of the great apes have babies called infants.

2. Baby elephants often suck on their trunk for comfort.

3. A baby hare is called a leveret.

4. Baby camels are born without humps.

5. They can run within hours of birth.

6. Shark fetuses will kill and eat other fetuses by the score in order to increase their chances of making it to birth.

7. A baby porcupine is a porcupette.

8. Only about one out of a thousand baby sea turtles survives after hatching.

9. A baby alpaca, llama, vicuña, or guanaco is a cria.

10. While play fighting, sometimes male puppies let female puppies win on purpose.

11. A baby swan is a cygnet or a flapper.

12. Kangaroos, koalas, opossums, wallabies, and wombats have babies called joeys.

13. When a baby joey is too big to fit in the pouch, a koala mom will carry her baby on her back.

14. Baby goats are called kids. Momma goats are called nannies.

15. Baby pandas weigh about as much as a cup of tea.

16. A young eel is called an elver.

17. Baby cheetahs are born without spots and develop them later.

18. In addition to "piglet," baby hogs and boars can also correctly be referred to as shoats (newly weaned pigs) or a farrow (a collective term for a group of young pigs).

19. Baby cougars are born with spots but lose them as they age.

20. A baby goose is a gosling.

21. Fox kits are born blind, deaf, and toothless, but mature quickly.

22. The name for a baby oyster is a spat.

23. Okapi calves can walk 30 minutes after birth, but they don't poop until they are between four and eight weeks old.

24. When a giraffe is ready to give birth after a 15 month gestation period, she'll often return to the place she was born.

25. Giraffes give birth standing up, so the baby (calf) falls from five or six feet to the ground.

26. Baby giraffes have little horns called "ossicones" just like adults, but they are folded down inside the womb.

27. They stand up straight a few hours after birth.

28. A baby pigeon is a squab or squeaker.

29. A baby echidna or platypus is called a puggle.

30. Baby fish are initially called larvae, then fry, and finally fingerlings before they mature into adult fish.

31. A newborn turkey chick has to be taught to eat, or it will starve.

15 Humdinger Hummingbird Facts

1. There are about 330 different species of hummingbirds.

2. Most live in Central and South America.

3. Due to their small body size and lack of insulation, hummingbirds lose body heat rapidly.

4. To meet their energy demands, they enter torpor (a state similar to hibernation), during which they lower their metabolic rate.

5. During torpor, the hummingbird drops its body temperature by 30° F to 40° F, and it lowers its heart rate from more than 1,200 beats per minute to as few as 50.

6. Excluding insects, hummingbirds have the fastest metabolism of any animal.

7. To provide energy for flying, hummingbirds must consume up to three times their body weight in food each day.

8. Unlike other birds, a hummingbird can rotate its wings in a circle.

9. A hummingbird can also hover in one spot; fly up, down, sideways, and even upside down (for short distances); and is the only bird that can fly backward.

10. The wing muscles of a hummingbird account for 25 to 30 percent of its total body weight, making it well adapted to flight.

11. However, the hummingbird has poorly developed feet and cannot walk.

12. Like bees, hummingbirds carry pollen from one plant to another while they are feeding, playing an important role in plant pollination.

13. Each hummingbird can visit between 1,000 and 2,000 blossoms every day.

14. The smallest bird on Earth is the bee hummingbird (*Calypte helenae*), native to Cuba.

15. It can comfortably perch on the eraser of a pencil.

29 Big Cat Facts

1. The "big cats" include the lion, tiger, jaguar, and leopard—all members of the genus *panthera*.

2. The snow leopard has been included in the genus *panthera* since 2008, but there is some controversy over moving them from their own genus, *uncia*.

3. Not all big cats have distinct larynx characteristics that allow them to roar.

4. Big cats are divided into two groups—those that roar, such as tigers and African lions, and those that purr.

5. Mountain lions purr, hiss, scream, and snarl, but they cannot roar.

6. While cheetahs can't roar, they can purr while inhaling and exhaling. Cheetahs have more in common with their domesticated small cousins.

7. The cheetah is the world's fastest land animal.

8. It can reach speeds of 70 miles per hour in under 3 seconds. Cheetahs can only maintain this pace for around 20 seconds.

9. Tigers are the most powerful cats in the world.

10. A Siberian tiger can grow up to 13 feet long and weigh as much as 660 pounds.

11. If you shaved a tiger, you'd discover striped skin under its fur.

12. A fully-grown male tiger can leap 33 feet in one jump.

13. Though they spend most of their time on land, tigers are one of the few cats that love water.

14. The white tiger is a rare variety of Bengal tiger with white fur and black stripes.

15. A tiger can tell if it's in another tiger's territory by sniffing trees marked with urine.

16. Leopards are the only big cats that regularly climb trees.

17. A leopard can easily drag prey twice its own body weight up to the branches of a tree.

18. Leopards eat everything from dung beetles to antelopes and young giraffes.

19. Black panthers are actually leopards with dark coats.

20. Jaguars are the largest cats in the Americas.

21. A jaguar's roar sounds more like a cough.

22. All lions live in groups called prides, usually consisting of one or two adult males, plus six to eight females and their cubs.

23. Because they are smaller, quicker, and more agile than males, female lions do the hunting.

24. While the lionesses are at work, the males patrol the area and protect the pride from predators.

25. Even though female lions do the hunting, male lions often get to eat first.

26. The African lion has the largest larynx and the loudest roar—it can be heard from five miles away.

27. The serval has the longest legs relative to its body of any cat.

28. A serval uses its large ears to listen for prey— then pounce.

29. Cheetahs will use their tails to help turn when they are running at full speed.

43 Marine Mammal Facts

1. Marine mammals are warm-blooded animals that live in or very near the ocean, breathe air, give live birth, and nurse their young.

2. Whales, dolphins, and porpoises are classified in the cetacean group of marine mammal.

3. There are over 70 different species of cetacean.

4. There are two suborders of whales—baleen whales (mysticetes) and toothed whales (odontocetes).

5. Baleen whales (called mysticeti) filter their food through huge comb-like plates made of flexible keratin.

6. Baleen whales include humpback whales, blue whales, North Atlantic right whales, and bowhead whales.

7. A blue whale can eat up to four tons of krill in a day.

8. The call of the blue whale registers an incredible 188 decibels, making it the loudest animal on Earth.

9. The blue whale is the largest animal that has ever lived (yes, including dinosaurs).

10. A newborn blue whale calf is almost as long as a school bus.

11. Male humpback whales sing complex songs in winter breeding areas that can last up to 20 minutes and be heard miles away.

12. Bowhead whales have massive skulls that can be over 16.5 feet long—or about 30 to 40 percent of their entire body length—that they use to break through ice.

13. When a whale surfaces after a dive, it exhales air from its lungs through its blowhole.

14. As the warmed air hits the colder ocean air, it condenses into droplets, making it appear that the whale is spouting water.

15. Unsurprisingly, toothed whales (called odonotoceti) have teeth that they use to sense, capture, and/or eat prey.

16. Whales in this category include narwhals, belugas, and sperm whales.

17. When hunting squid, a sperm whale may spend as much as an hour on a dive to depths of more than 3,000 feet, where the pressure is more than 1,400 pounds per square inch.

18. Young sperm whales often form coed schools that gradually split up as dominant males drive off smaller ones until just one male is left with a harem of up to 25 females.

19. Some sperm whales swim enough miles during their lifetime to circle Earth several times.

20. Most marine mammals produce urine that is saltier than seawater.

21. Beluga whales have a five-inch-thick layer of blubber and dorsal ridge that help them navigate through harsh, icy waters.

22. Beluga whales' complex communication repertoire of whistles, clicks, and chirps has prompted the nickname "canaries of the sea."

23. Whales can't swim backwards.

24. The killer whale, or orca, will actually beach itself to feast on baby seals. It then worms its way back to the water.

25. Don't let the "killer whale" name fool you. The orca is really a dolphin, the largest member of the dolphin family.

26. More than 30 species of dolphin inhabit every ocean in the world—even some rivers.

27. Each dolphin has a unique whistle it uses to identify itself.

28. Bottlenose dolphins surface two or three times per minute for air.

29. Right whale dolphins are the only members of the dolphin family without dorsal fins.

30. Seals, sea lions, and walruses are all grouped together in the scientific order Pinnipedia, which is Latin for "fin-footed."

31. All pinnipeds have four "fins" or flippers.

32. Pinnipeds spend most of their lives swimming and eating in water and come onto land or ice to give birth to their young, rest, and molt.

33. Phocidae ("earless" seals) have tiny ear holes but no external ear flaps.

34. Walruses in the Pacific migrate with moving ice, climbing atop it to rest and give birth.

35. Manatees and dugongs are sirenians.

36. Like cetaceans, sirenians spend their whole lives in water.

37. Sirenians are the only group of marine mammal that eat no meat.

38. Polar bears and sea otters are marine fissipeds, "split-footed" members of the order Carnivora.

39. Although skilled swimmers, polar bears are the marine mammal least adapted to aquatic existence.

40. They rest, mate, give birth, and nurse their young on the ice. They are particularly vulnerable to reductions in sea ice.

41. Polar bears use a nose-to-nose greeting when asking another bear for something.

42. Sea otters hold hands while sleeping to keep from drifting apart.

43. Sea otters have a pouch under their forearm to store food and their favorite rocks.

Chapter 4

HEALTH AND THE HUMAN BODY

18 Facts About Cells

1. The exact cell count in the human body is unknown, and figuring it out is tricky.

2. One popular estimate is 30 trillion cells; another says 37 trillion.

3. About 300 million cells die every minute.

4. Red blood cells are the most common type of cell.

5. More than 80 percent of the cells in the human body are erythrocytes, or red blood cells.

6. Cells of the same type group together to form tissue.

7. Fifteen million blood cells are produced and destroyed in the human body every second.

8. Mitochondria use oxygen to produce energy for the cell.

9. The longest cells in the human body are the motor neurons.

10. They can be up to four-and-a-half-feet long and run from the lower spinal cord to the big toe.

11. The longest-living cells in the body are brain cells, which can live a human's entire lifetime.

12. The cells that control smell and taste are the only sensory cells that are replaced during a person's life span.

13. There are probably about as many bacteria and microbe cells in our body as there are human cells.

14. For a long time people thought bacteria might outnumber human cells by 10 to 1, but later revisions brought the estimate down.

15. Fat cells are called adipocytes or lipocytes.

16. Epithelial cells line your skin, the outsides of your organs, and the insides of some organs.

17. They protect your body and organs.

18. Stem cells generate other types of cells.

21 Heart & Circulatory System Facts

1. The circulatory system moves blood throughout the body.

2. Its main components are the heart, blood vessels, and blood.

3. Laid end to end, there are about 60,000 miles of blood vessels in the human body.

4. The hard-working heart pumps about 2,000 gallons of blood through those vessels every day.

5. The human heart creates enough pressure while pumping to squirt blood 30 feet.

6. Blood consists of red blood cells, white blood cells, plasma, and platelets.

7. Newborn babies only have about 1 cup of blood.

8. The average adult has about 5 liters of blood.

9. The right side of your heart pumps blood into your lungs.

10. The left side of your heart pumps blood back through your body.

11. Arteries carry oxygenated blood from the heart into smaller sections called arterioles and then into even smaller sections called capillaries.

12. Nutrients carried in the blood diffuse into body tissues from the capillaries.

13. The word capillary comes from the Latin *capillus*, hair.

14. Veins carry the deoxygenated blood from body tissues back to the heart.

15. The human heart beats over 100,000 times per day.

16. The average woman's heart is 8 beats per minute faster than a man's heart.

17. The first open-heart surgery was performed in 1893 by Daniel Hale Williams, one of the few Black cardiologists in the U.S. at the time.

18. The human heart weighs less than 1 pound.

19. A man's heart, on average, is 2 ounces heavier than a woman's heart.

20. Approximately 20 percent of the oxygenated blood flowing from the heart is pumped to the brain.

21. Other than the cornea, every cell in the body gets blood from the heart.

25 Facts About Your Head

1. The human head is one-quarter of our total length at birth.

2. But the head is only one-eighth of our total length by the time we reach adulthood.

3. If your mouth was completely dry, you would not be able to distinguish the taste of anything.

4. Your nose is not as sensitive as a dog's, but it can remember 50,000 different scents.

5. The air from a human sneeze can travel at speeds of 100 miles per hour or more.

6. Each of your nostrils registers smell differently.

7. The right nostril detects the more pleasant smells, but the left one is more accurate.

8. Women are usually better than men at identifying specific smells.

9. Most people only have a 50 percent success rate at detecting a single drop of perfume in a three-room apartment.

10. Don't stick out your tongue if you want to hide your identity!

11. Similar to fingerprints, everyone also has a unique tongue print.

12. The human eye blinks an average of 3.7 million times per year.

13. The pupil of the eye expands as much as 45 percent when a person looks at something pleasing.

14. The retina of the eye is the only part of the central nervous system that can be seen from the outside of the body— but you have to look directly through the pupil to see it.

15. When we blink (about 20,000 times a day) our brain keeps things illuminated so the world doesn't go dark each time.

16. It takes six muscles to move each of your eyeballs.

17. There's a facial muscle called the levator labii superioris alaeque nasi muscle.

18. It's the longest muscle name.

19. Perhaps because of its long name, it's also known as the Elvis muscle, because it's used to raise your upper lip in the way Elvis famously did.

20. The body's smallest bones are found in the middle ear.

21. Changes to the cartilage in your ears means that they lengthen throughout the course of your lifetime.

22. You have rocks in your body!

23. Your inner ear has tiny crystals that help you maintain balance.

24. The average skull is about 0.26 to 0.27 inches thick.

25. Women's skulls are slighter thicker on average than men's.

22 Sweaty Facts

1. Sweat enables us to cool off when the exterior temperature rises (due to changes in the weather) or when our interior temperatures rise (due to exercise, anxiety, or illness).

2. Sweat is one of the mechanisms that our bodies use to keep us at a steady—and healthy—98.6 degrees Fahrenheit.

3. Humans have about 2.6 million sweat glands.

4. Not all of these glands produce the same kind of sweat.

5. Sweat has two distinct sources: eccrine and apocrine glands.

6. Eccrine glands exist all over the body and are active from birth.

7. They constantly release a salty, nearly odorless fluid onto the skin.

8. Apocrine glands are concentrated in the armpits, on the soles of the feet, in the palms of the hands, and in the groin.

9. They become active during puberty. Yes, puberty and perspiration go hand in hand.

10. Apocrine glands don't secrete liquid directly onto the skin.

11. Instead, each gland empties into a hair follicle.

12. When a person is under emotional or physical stress, the tiny muscle around the hair follicle contracts, pushing the liquid onto the skin, where it becomes sweat.

13. Apocrine glands carry lipids and proteins, as well as water and salt.

14. When these substances mix with the sebaceous oils in the hair follicles and then meet the bacteria on the skin, well, that's when you begin to hold your nose.

15. Apocrine sweat has been found to contain androsterone pheromones, those mysterious musky odors that are responsible for sexual arousal.

16. Deodorants are based on mildly acidic compounds that dry the skin before the odor starts.

17. Antiperspirants, another popular option, actually block sweat with aluminum salts.

18. Excessive sweating is officially known as hyperhidrosis.

19. Lack of sweat is called anhidrosis.

20. Both hyperhidrosis and anhidrosis are genuine medical conditions with serious complications.

21. Fortunately, both are treatable.

22. For most of us, dealing with sweat is fairly simple: Take a shower and wear loose, absorbent clothing.

8 Most Common Blood Types

Scientists have discovered eight major blood types. See the major types, and how common they are in the U.S. population.

  1. O+: 37.4 percent

  2. A+: 35.7 percent

3. B+: 8.5 percent

4. O-: 6.6 percent

5. A-: 6.3 percent

6. AB+: 3.4 percent

7. B-: 1.5 percent

8. AB-: 0.6 percent

37 Blood Type Facts

1. There are four major blood groups (A, B, AB, O) determined by the presence or absence of two antigens, A and B, on the surface of red blood cells.

2. Group A has only the A antigen on red blood cells (and B antibody in the plasma).

3. Group B has only the B antigen on red blood cells (and A antibody in the plasma).

4. Group AB has both A and B antigens on red blood cells (but neither A nor B antibody in the plasma).

5. Group O has neither A nor B antigens on red blood cells (but both A and B antibody are in the plasma).

6. Like eye color, blood type is inherited.

7. Whether your blood group is A, B, AB, or O is based on the blood types of your biological parents.

8. In addition to the A and B antigens, there is a protein called the Rh factor, which can be either present (+) or absent (–), creating the eight most common blood types: A+, A-, B+, B-, AB+, AB-, O+, and O-.

9. The most common blood type is O+, which occurs in about 37 percent of the U.S. population.

10. About 53 percent of Latinos have type O+ blood, as do 47 percent of African Americans, 39 percent of Asian Americans, and 37 percent of Caucasians.

11. O+ blood can be given to a person with A+, B+, AB+, or O+ blood.

12. A person with O+ blood can receive blood from O+ or O- donors.

13. A+ blood occurs in about 35 percent of the population.

14. Type A+ occurs in about 33 percent of Caucasians, 29 percent of Latinos, 27 percent of Asian Americans, and 24 percent of African Americans.

15. A person with A+ blood can receive A+, A-, O+, or O- blood.

16. However, A+ blood can be given only to a person with the A+ or AB+ blood types.

17. B+ blood occurs in about 8 percent of the U.S. population.

18. About 25 percent of Asians have B+ blood, as do 18 percent of African Americans and 9 percent of Latinos and Caucasians.

19. B+ blood can be given only to those with either AB+ or B+ blood.

20. This blood type can receive blood from B+, B-, O+, or O- donors.

21. O- is considered the universal donor because it can be given to anyone, regardless of blood type.

22. However, a person with O- blood can receive blood only from other O- donors.

23. About 8 percent of Caucasians have type O-, as do 4 percent of African Americans and Latinos, and 1 percent of Asians.

24. A- blood can be given to a person with AB-, A-, AB+, or A+.

25. A person with type A- can only receive blood from O- or A- donors.

26. Approximately 7 percent of Caucasians have type A-, as do 2 percent of Latinos and African Americans, and 0.5 percent of Asians.

27. AB+ is considered a universal receiver because people with this blood type can receive blood of any type.

28. But AB+ blood can only be given to a person who also has AB+.

29. Approximately 7 percent of Asian Americans have type AB+, as do 4 percent of African Americans, and 2 percent of Latinos and Caucasians.

30. B- blood can be given to those with B-, AB-, B+, or AB+ blood.

31. A person with B- blood can receive blood from O- or B- blood types.

32. About 2 percent of Caucasians, 1 percent of African Americans and Latinos, and 0.4 percent of Asians have type B- blood.

33. AB- is the rarest of the eight common blood types, occurring in about 0.6 percent of the U.S. population.

34. People with type AB- can give blood to AB+ or AB- blood types.

35. People with type AB- must receive blood from O-, A-, B-, and AB- blood types.

36. AB- blood occurs in about 1 percent of Caucasians, 0.3 percent of African Americans, 0.2 percent of Latinos, and 0.1 percent of Asian Americans.

37. The universal plasma donor has type AB blood.

26 History of the Lobotomy Facts

1. There's a reason why lobotomies have taken a place in the Health Care Hall of Shame.

2. A lobotomy is a surgical procedure that severs the paths of communication between the prefrontal lobe and the rest of the brain.

3. This prefrontal lobe—the part of the brain closest to the forehead—is a structure that appears to have great influence on personality and initiative.

4. So the obvious question is: Who thought it would be a good idea to disconnect it?

5. It started in 1890, when German researcher Friederich Golz removed portions of his dog's brain.

6. He noticed afterward that the dog was slightly more mellow—and the lobotomy was born.

7. The first lobotomies performed on humans took place in Switzerland two years later.

8. The six patients who were chosen all suffered from schizophrenia.

9. While some did show post-op improvement, two others died.

10. Apparently this was a time when an experimental medical procedure that killed 33 percent of its subjects was considered a success.

11. Despite these grisly results, lobotomies became more commonplace.

12. One early proponent of the surgery even received a Nobel Prize.

13. The most notorious practitioner of the lobotomy was American physician Walter Freeman.

14. From the 1930s to the 1960s, Freeman performed the procedure on more than 3,000 patients.

15. Rosemary Kennedy, the sister of President John F. Kennedy, was one such patient.

16. Freeman pioneered a surgical method in which a metal rod (known colloquially as an "ice pick") was inserted into the eye socket, driven up into the brain, and hammered home.

17. This is known as a transorbital lobotomy.

18. Freeman and other doctors in the U.S. lobotomized an estimated 40,000 patients before an ethical outcry over the procedure prevailed in the 1950s.

19. Although the mortality rate had improved since the early trials, it turned out that the ratio of success to failure was not much higher.

20. A third of the patients got better, a third stayed the same, and a third became much worse.

21. The practice had generally ceased in the United States by the early 1970s.

22. It is now illegal in some states.

23. Lobotomies were performed only on patients with extreme psychological impairments, after no other treatment proved successful.

24. The frontal lobe of the brain is involved in reasoning, emotion, and personality, and disconnecting it can have a powerful effect on a person's behavior.

25. Unfortunately, the changes that a lobotomy causes are unpredictable and often negative.

26. Today, there are far more precise and far less destructive manners of affecting the brain through antipsychotic drugs and other pharmaceuticals.

21 Facts About Trepanation

1. Trepanation (also known as "trephination") is the practice of boring into the skull and removing a piece of bone, thereby leaving a hole.

2. It is derived from the Greek word *trypanon*, meaning "to bore."

3. This practice dates back 7,000 years and is still practiced today.

4. It was performed by the ancient Greeks, Romans, and Egyptians, among others.

5. Hippocrates, considered the father of medicine, indicated that the Greeks might have used trepanation to treat head injuries.

6. However, evidence of trepanning without accompanying head trauma has been found in other civilizations.

7. Speculation abounds as to its exact purpose.

8. Since the head was considered a barometer for a person's behavior, one theory is that trepanation was used as a way to treat headaches, depression, and other conditions that had no outward trauma signs.

9. In trepanning, the Greeks used an instrument called a *terebra*, an extremely sharp piece of wood with another piece of wood mounted crossways on it as a handle and attached by a thong.

10. The handle was twisted until the thong was extremely tight.

11. When released, the thong unwound, which spun the sharp piece of wood around and drove it into the skull like a drill.

12. Although it's possible that the *terebra* was used for a single hole, it is more likely that it was used to make a circular pattern of multiple small holes, thereby making it easier to remove a large piece of bone.

13. The Incas were also adept at trepanation.

14. The procedure was performed using a ceremonial tumi knife made of flint or copper.

15. The surgeon held the patient's head between his knees and rubbed the tumi blade back and forth along the surface of the skull to create four incisions in a criss-cross pattern.

16. When the incisions were sufficiently deep, the square-shaped piece of bone in the center was pulled out.

17. Doctors still use this procedure, only now it's called a craniotomy.

18. It still involves removing a piece of skull to get to the underlying tissue.

19. The bone is replaced when the procedure is done.

20. If it is not replaced, the operation is called a craniectomy.

21. That procedure is used in many different circumstances, such as for treating a tumor or infection.

28 Medical Slang Terms Decoded

1. **Appy:** a person's appendix or a patient with appendicitis

2. **Baby Catcher:** an obstetrician

3. **Bagging:** manually helping a patient breathe using a squeeze bag attached to a mask that covers the face

4. **Banana:** a person with jaundice (yellowing of the skin and eyes)

5. **Blood Suckers/Leeches:** those who take blood samples, such as laboratory technicians

6. **Bounceback:** a patient who returns to the emergency department with the same complaints shortly after being released

7. **Bury the Hatchet:** accidentally leaving a surgical instrument inside a patient

8. **CBC:** complete blood count; an all-purpose blood test used to diagnose different illnesses and conditions

9. **Code Brown:** a patient who has lost control of his or her bowels

10. **Code Yellow:** a patient who has lost control of his or her bladder

11. **Crook-U:** similar to the ICU or PICU, but referring to a prison ward in the hospital

12. **DNR:** do not resuscitate; a written request made by terminally ill or elderly patients who do not want extraordinary efforts made if they go into cardiac arrest, a coma, etc.

13. **Doc in a Box:** a small health-care center, usually with high staff turnover

14. **FLK:** funny-looking kid

15. **Foley:** a catheter used to drain the bladder of urine

16. **Freud Squad:** the psychiatry department

17. **Gas Passer:** an anesthesiologist

18. **GSW:** gunshot wound

19. **MI:** myocardial infarction; a heart attack

20. **M & Ms:** mortality and morbidity conferences where doctors and other health-care professionals discuss mistakes and patient deaths

21. **MVA:** motor vehicle accident

22. **O Sign:** an unconscious patient whose mouth is open

23. **Q Sign:** an unconscious patient whose mouth is open and tongue is hanging out

24. **Rear Admiral:** a proctologist

25. **Shotgunning:** ordering a wide variety of tests in the hope that one will show what's wrong with a patient

26. **Stat:** from the Latin *statinum*, meaning immediately

27. **Tox Screen:** testing the blood for the level and type of drugs in a patient's system

28. **UBI:** unexplained beer injury; a patient who appears in the ER with an injury sustained while intoxicated that he or she can't explain

13 Facts About Taste Buds

1. Babies are born with taste buds on the insides of their cheeks.

2. Overall babies have more taste buds than adults, but they lose them as they grow older.

3. Adults have, on average, around 10,000 taste buds.

4. An elderly person might have only 5,000.

5. One in four people is a "supertaster" and has more taste buds than the average person—more than 1,000 per square inch.

6. Twenty-five percent of humans are "nontasters" and have fewer taste buds than other people their age—only about 40 per square inch.

7. A taste bud is 30 to 60 microns (slightly more than 1/1000 inch) in diameter.

8. Taste buds are not just for tongues—they also cover the back of the throat and the roof of the mouth.

9. Attached to each taste bud are microscopic hairs called microvilli.

10. Taste buds are regrown every two weeks.

11. About 75 percent of what we think we taste is actually coming from our sense of smell.

12. Along with sweet, salty, sour, and bitter, there is a fifth taste, called umami, which describes the savory taste of foods such as meat, cheese, and soy sauce.

13. The absolute threshold of taste is one teaspoon of sugar mixed into two gallons of water—most people can only taste it 50 percent of the time.

62 Brain & Nervous System Facts

1. The weight of the human brain triples during the first year of life, going from about 10 ounces to about 30 ounces.

2. At birth, babies have about 100 billion brain cells, but most of their neurons aren't yet connected.

3. That process is complete by age 3.

4. Only 4 weeks after conception, a human embryo's brain is already developing at an astonishing pace.

5. During this stage of early fetal development, neurons are forming at a rate of 250,000 per minute!

6. The average adult human brain weighs about 3 pounds.

7. It makes up about 2 percent of the total body weight.

8. After you're 30 years old, the brain shrinks a quarter of a percent in mass each year.

9. From early childhood through puberty, synapses in the human neocortex are lost at a rate of 100,000 per second.

10. Fevers are controlled by the part of the brain called the hypothalamus.

11. Temperatures greater than 109 degrees can be fatal.

12. Located in the lower back portion of the brain, the cerebellum controls such things as posture, walking, and coordination.

13. Scientists also think the cerebellum plays a role in the way scents are processed.

14. The neurons in the human brain allow information to travel at speeds up to 268 miles per hour.

15. The brain itself does not feel pain, so neurosurgeons can perform brain operations while patients are awake.

16. The brain contains on average 86 billion neurons. (The common figure once cited was 100 billion neurons, but more specific tests lowered the estimate.)

17. There are 1,000 to 10,000 synapses for each neuron in the brain.

18. An active brain produces new dendrites, which are the connections between nerve cells that allow them to communicate with one another.

18. Fat makes up about 10 percent of your brain.

19. A fatty substance called myelin wraps and insulates a number of your brain's neurons.

20. Without the insulator myelin, which increases brain efficiency, the brain would be ten times bigger and each person would have to eat ten times as much food to provide enough energy for the brain to function.

21. The frontal lobes of your brain create feelings of self-awareness.

22. Evidence suggests that children develop self-awareness at around 18 months of age.

23. The outer part of your brain—the cortex—is split into right and left hemispheres.

24. They are connected by a bundle of 50 million neurons.

25. Brain scans of cab drivers in London showed that their hippocampi—the part of the brain that helps us navigate—was larger than those of other people.

26. A piece of a human brain the size of a grain of sand contains 100,000 neurons and 1 billion synapses, all "talking" to one another.

27. The first cervical dorsal spinal nerve and dorsal root ganglia, which help bring sensory information into the brain and spinal cord, are missing in 50 percent of all people.

28. The part of your brain that keeps risky behavior in check isn't fully formed until 25 years of age.

29. Dates, statistics, and other factual memories (such as trivia) are stored in the front left side of the brain.

30. The human brain is about 75 percent water.

31. Your brain uses fatty acids from fats to create the specialized cells that allow you to think and feel.

32. Your amygdala is responsible for generating negative emotions such as anger, sadness, fear, and disgust.

33. Working on non-emotional mental tasks inhibits the amygdala, which is why keeping yourself busy can cheer you up when you're feeling down.

34. Alcohol weakens connections between neurons and makes new cells grow less quickly, which interfere with brain activity and causes serious damage.

35. Scientists have found that it is impossible to learn something well enough to create a "permanent" memory.

36. All memories have a limited lifetime.

37. The first person to record electrical activity in the brain was Richard Caton.

38. He accomplished this feat in 1875.

39. An estimated 3 million Americans stutter, mostly children under age six.

40. By adulthood, fewer than 1 percent stutter.

41. Some scientists believe that stutterers have neurological differences from non-stutterers.

42. The human spinal cord houses about 1 billion neurons.

43. An image of a single item, such as a house or a face, activates at least 30 million neurons in the visual cortex of the brain.

44. The average person produces between 14 and 17 fluid ounces of cerebrospinal fluid every day.

45. Wearing a helmet when riding a bicycle can reduce the risk for brain injury by up to 88 percent.

46. The brain uses a whopping 17 percent of the body's energy.

47. The stress caused by frequent jetlag and changing work hours (like those experienced by airline employees and shift workers) can damage memory and the temporal lobe of the brain.

48. People who have damage in their brain's frontal lobe may not be able to tell when the punch line of a joke is funny.

49. Your brain can tell the difference between your own touch and someone else's—that's why you can't tickle yourself.

50. The amygdala allows you to read people's faces to determine how they are feeling.

51. Male and female brains have different reactions to bodily pain.

52. An ancient Greek doctor named Alcmaeon was the first to conclude—in 450 B.C.!—that the brain, not the heart, is the origin of thoughts and feelings.

53. Scientists have found that the planum temporale is the area of the brain responsible for giving musicians perfect pitch.

54. Over time, human brains have increased in size—at a rate of about 0.5 percent per decade.

55. Since an individual brain neuron measures only four microns in thickness, about 30,000 neurons would fit on the head of a pin.

56. The group of spinal nerves at the lower end of the spinal cord is called the cauda equina.

57. The name is perfectly descriptive: It's Latin for "tail of a horse."

58. People with the rare disorder agnosia (damage to areas of the occipital or parietal brain lobes) can't recognize and identify objects, and may not know whether a person's face is familiar to them.

59. Prolonged stress can kill cells in the hippocampus, the part of your brain that's critical for memory.

60. Thankfully, we're able to grow new neurons in this area again, even as adults.

61. Your brain is swaddled in several layers of membranes called meninges.

62. The fluid between these layers produces a water cushion that protects your brain if you bump your head.

24 Facts About Skin & Nails

1. Skin is your largest body organ, accounting for about 15 percent of your body weight.

2. Your skin covers an area of about 21 square feet (2 square m).

3. Humans shed about 600,000 particles of skin every hour.

4. That works out to about 1.5 pounds each year, so the average person will lose around 105 pounds of skin by age 70.

5. New skin cells take about a month to reach the surface as the cells above them die off.

6. Every square inch of human skin has about 32 million bacteria on it, but fortunately, the vast majority of them are harmless.

7. A pair of feet have 500,000 sweat glands and can produce more than a pint of sweat a day.

8. If you're clipping your fingernails more often than your toenails, that's only natural.

9. The nails that get the most exposure and are used most frequently grow the fastest.

10. Fingernails grow fastest on the hand that you write with and on the longest fingers.

11. On average, nails grow about one-tenth of an inch each month.

12. Fingerprints are formed while a fetus is growing and are the result of DNA and environmental influences in the womb.

13. Factors such as contact with amniotic fluid and the pressure of bone growth affect the unique patterns.

14. By the second trimester of pregnancy, the ridges and loops in our digits are permanently etched into our skin.

15. Fingerprint identification is far from an exact "science."

16. Analysts look for points of similarity, but there are no universal standards, and no research dictates the number of points that establish a match with certainty.

17. Goosebumps happen when the muscles around the follicles of your body hair contract.

18. In animals with more fur, this reflex helps keep the body warm.

19. Your legs contain fewer oil glands than any other part of your body.

20. You have between 2 and 4 million eccrine sweat glands—those are the glands that are found all over your body.

21. Receptors—cells that communicate with sensory nerves—differ in the sizes of their receptive fields.

22. The receptors in your fingertips have smaller fields than the receptors on other parts of your body, making them better able to distinguish touches.

23. Your hands have about 17,000 touch receptors, concentrated in your fingertips.

24. Your eyelids have the thinnest skin of your body.

13 Medical Device Facts

1. A spirometer measures how much air your lungs take in and expel.

2. An autoclave uses heat to sterilize medical equipment.

3. A portable Automated External Defibrillator, or AED, delivers an electric shock to restore heart rhythm.

4. The device that doctors use to look inside your ear is called an otoscope.

5. Light and a magnifying lens make it easier for the doctor to see ear problems.

6. Built along the same line, a rhinoscope is used for examinations of the nose.

7. An ophthalmoscope is used to examine the eye.

8. The fancy name for a blood pressure monitor is a sphygmomanometer.

9. It's from Greek words for heartbeat and measurement.

10. The reflex hammer used to test reflexes comes in several models, including the Queen Square, the Babinski, the Buck, the Berliner, and the Stookey.

11. Electrocardiograph machines test the electrical function of the heart.

12. An ECG (or EKG) can detect heart rhythm abnormalities and blockages.

13. A tympanometer tests middle-ear function.

56 History of Anesthesia Facts

1. Today, it's nearly unimaginable to consider having surgery without the aid of anesthesia.

2. Yet for most of human history, people had no surefire way to guarantee pain-free surgery.

3. Humans have long tried to find ways to dull pain.

4. Alcohol was probably the earliest method, used in ancient Mesopotamia, where the Sumerians also cultivated opium poppy.

5. The opium latex obtained from the poppy is about 12 percent morphine, and was described by the Sumerians as *hul gil*, or "plant of joy."

6. The Babylonians soon became aware of opium's numbing effects, and the plant, along with their empire, next spread to Persia and Egypt.

7. The ancient Egyptians also made crude sedatives out of the mandrake fruit, which contains psychoactive agents that can cause hallucinations.

8. In India and China, cannabis incense was used to promote a feeling of relaxation, sometimes along with a cup of wine.

9. According to eyewitness reports, Bian Que, an early Chinese physician who practiced in the mid-300s B.C.,

was said to have used a "toxic drink" to induce a coma-tose-like state in two patients for several days.

10. Five centuries later, Chinese physician Hua Tuo is said to have used a combination of sedating herbs he called *mafeisan*, which would be dissolved in wine for a patient to drink before surgery.

11. According to surviving records, Hua Tuo performed many surgeries with the help of this mixture.

12. But its exact recipe was lost when the physician burned his own notes shortly before his death.

13. During the Middle Ages, Arabic and Persian physicians discovered that anesthetics could be inhaled as well as ingested.

14. Persian physician Ibn Sina, also known by the Latinized name Avicenna, described in the 1025 encyclopedia *The Canon of Medicine* a method of holding a sponge infused with narcotics beneath a patient's nose.

15. By the 13th century, the English were using a mixture of bile, opium, bryony, henbane, hemlock, lettuce, and vinegar called *dwale*, as a sedative.

16. This potion was not administered by physicians, who warned against its use, possibly because some of the ingredients can be poisonous.

17. Recipes to make the mixture were found in remedy books that ordinary housewives could tuck right next to their cookbooks.

18. By 1525, Swiss physician Paracelsus had discovered that the compound diethyl ether had analgesic properties.

19. But it would take more than two centuries before physicians would start to consider using it as an anesthetic.

20. These practices all had various levels of success and plenty of risks, but nothing was foolproof when it came to surgical anesthetic.

21. In fact, by the turn of the 19th century, surgery was performed only as a last resort—usually with very little, if any, anesthetic.

22. Not surprisingly, this caused immense fear and pain for patients, and extreme anxiety for the doctors who had to operate on them.

23. Thankfully, the discovery of gases like oxygen, ammonia, and nitrous oxide in the late 1700s gave scientists new ideas, and physicians and patients some new options.

24. English physician Thomas Beddoes was one of the first to take notice of the new discovery of gases.

25. Beddoes founded a medical research facility called the Pneumatic Institution in 1798.

26. His aim was to find therapeutic ways to use different gases.

27. Beddoes hoped to treat breathing issues like asthma and tuberculosis.

28. While working at the Pneumatic Institution, chemist Humphry Davy (who would later go on to invent the field of electrochemistry) discovered that nitrous oxide had analgesic effects.

29. He also noted that it produced a feeling of euphoria, which prompted him to call it "laughing gas."

30. Davy suggested that nitrous oxide gas should be used during surgical procedures.

31. This idea was not immediately acted upon, perhaps because Davy himself was not a physician.

32. In 1813, Davy was joined by a new assistant, Michael Faraday (now best known for his work with electromagnetism).

33. Faraday began studying the inhalation of ether and found it to also have analgesic effects, as well as having the ability to cause sedation.

34. Even after he published his findings in 1818, ether was mostly ignored by surgeons, too.

35. But strangely, in the United States, these gases found another use, when people realized they could be inhaled recreationally at "laughing gas parties" and "ether frolics."

36. By the mid-1800s, traveling lecturers and showmen would hold these unusual gatherings.

37. Members of the audience were encouraged to inhale nitrous oxide or ether while other audience members laughed at the mind-altering results.

38. One of the participants of these "ether frolics" was a young dentist named William Morton, who was intrigued by the analgesic potential of the gas.

39. Morton tested it on animals and then successfully used it for several patients.

40. After that, Morton was so confident in ether's ability to anesthetize that he offered to demonstrate his method to Dr. John Warren, the surgeon at Massachusetts General Hospital.

41. On October 16, 1846, before a large audience in what is now known as the "Ether Dome," Morton administered diethyl ether to a young patient named Edward Gilbert Abbott.

42. Warren then removed a tumor from Abbott's neck.

43. To the surprise of everyone in attendance, even the surgeon, Abbott appeared to remain comfortable during the entire procedure.

44. Afterwards, Abbott reported that he'd felt a scratching sensation, but no pain.

45. News of Abbott's surgery quickly spread around the world.

46. By December 1846, physicians in Great Britain were making use of inhalation anesthesia.

47. Scottish obstetrician James Young Simpson first used chloroform in 1847.

48. The use of ether and chloroform made surgery a less distressing venture for patient and surgeon alike.

49. The success of the compounds led to more anesthesia research.

50. The first intravenous anesthesia, sodium thiopental, debuted for human use in March 1934.

51. In the second half of the 20th century, Belgian doctor Paul Janssen synthesized more than 80 pharmaceutical compounds, including drugs used for anesthesia.

52. Ether and chloroform eventually fell out of favor as safer anesthesia alternatives became available.

53. Ether was highly flammable, so could not be used once electricity was used to monitor patients or when wounds were being cauterized.

54. Chloroform became associated with a high number of cardiac arrests.

55. Modern anesthesia is among the safest of all medical procedures.

56. Surprisingly, however, scientists are still uncertain exactly how anesthesia is able to affect our conscious minds.

18 Facts About Bones

1. An adult has fewer bones than a baby.

2. We start off life with 350 bones, but because bones fuse together during growth, we end up with only 206 as adults.

3. People stop growing taller when the growth plates at the ends of their bones fuse.

4. This tends to happen around 14 or 15 for girls, and is tied to the age at which they began menstruation.

5. Most boys reach their full height around age 16 to 18, but it can occur later.

6. While bones stop growing, bone cells regenerate.

7. More than a quarter of your bones are found in your hands and wrists: 27 bones in each.

8. Another quarter of your bones are located in your feet: 26 in each foot.

9. The largest bone of your foot, the heel bone, is called the calcaneus.

10. A child's bone heals more rapidly from a break than an adult's.

11. The hyoid bone in your throat isn't connected to a joint, unlike any other bone.

12. The femur is the longest, strongest, and largest bone in the body.

13. The tibia is named for its resemblance to a Latin pipe called a tibia.

14. The radius is named after its resemblance to the spoke of a wheel.

15. The fibula is named for its resemblance to a clasp in a brooch.

16. Your phalanges—or finger bones—are tied linguistically to the term phalanx, referring to an army formation.

17. The enamel on your teeth is stronger than your bones.

18. Your big toe is known as the hallux.

27 Digestive System Facts

1. In a lifetime, the average person produces about 25,000 quarts of saliva—enough to fill two swimming pools!

2. You get a new stomach lining every three to four days.

3. If you didn't, the strong acids your stomach uses to digest food would also digest your stomach.

4. Your stomach secretes about half a gallon of hydrochloric acid each day.

5. The small intestine is about four times as long as the average adult is tall.

6. Its loops would stretch out to 18 to 23 feet.

7. Your large intestine, though thicker than the small intestine, is actually shorter: only about 5 feet.

8. Your appendix is found right off your large intestine and is technically part of your gastrointestinal (GI) tract.

9. Scientists aren't entirely sure of the function of the appendix, and humans can easily live without it, but one hypothesis is that it stores "good" bacteria.

10. Food travels from the mouth, through the esophagus, and into the stomach in seven seconds.

11. Your esophagus is about 10 inches long.

12. The fancy name for the sound your digestive system makes when your stomach growls is borborygmus.

13. That noise is made by gas moving through your small intestine, not just your stomach, but "my small intestine is growling" doesn't have the same ring.

14. When you eat a meal, it takes anywhere from two to five days for it to work its way through your system.

15. The bulk of that time is spent in the large intestine.

16. Doctors can track how long it takes for food to make its way through your system—and where it might be getting stuck—with a bowel transit time test in which you swallow a pill with a wireless transmitter.

17. Different foods are digested at different rates.

18. Your body processes carbohydrates more quickly than proteins or fats, so fattier foods like meat takes longer to digest than fruits and vegetables.

19. That means that if you want to feel full for longer, a snack with a little bit of fat (cheese and crackers rather than just crackers) will help tide you over until dinner.

20. Some people naturally have more of the enzyme that helps you digest beans and other "gassy" vegetables.

22. Because the digestive system uses muscles, not gravity, to do its work, you could technically eat upside down.

23. In the 1820s, a fur trapper named Alexis St. Martin was accidentally shot and left with a hole in his side that meant his doctor William Beaumont could perform experiments to learn about stomachs and the digestive system.

24. Beaumont conducted experiments for a decade, treating St. Martin as a servant during that time, before St. Martin left the area and refused to return.

25. Each day, a healthy individual releases a minimum of 17 ounces of gas due to flatulence.

26. Most gas is composed of odorless hydrogen, nitrogen, and carbon dioxide.

27. In some humans—about 30 percent of the adult population—the digestive process also produces methane.

23 Facts About Hair

1. The average human head has 100,000 hair follicles, each of which is capable of producing 20 individual hairs during a person's lifetime.

2. Hair color helps determine how dense the hair on your head is.

3. Blondes (only natural ones, of course) have the densest hair, averaging 146,000 follicles.

4. People with black hair tend to have about 110,000 follicles.

5. Those with brown hair are about average with 100,000 follicles.

6. Redheads have the least dense hair, averaging about 86,000 follicles.

7. Each hair grows approximately 5 inches per year.

8. Nearly 80 percent of people in the United States say they spend more money on hair products than any other grooming goods.

9. Despite all the effort we put into coifing and pampering our locks, the hair we see is biologically dead.

10. Hair is alive only in the roots, which are fed by small blood vessels beneath the skin's surface.

11. Hair cells travel up the shaft and are eventually cut off from the blood supply that is their nourishment.

12. The cells die before being pushed out of the follicle onto the head—or back, arm, or anywhere else.

13. Hair is incredibly strong.

14. The average head of hair can support roughly 12 tons of weight—much more than the scalp it's attached to.

15. Hair grows in cycles.

16. During the first phase, hair is actively growing.

17. In the second phase, it rests in the follicle until it is pushed out of the root.

18. Healthy hair grows about 0.39 inches in a month.

19. Each hair is completely independent from the others.

20. When a hair falls out, another one may not grow directly in its place.

21. The average person loses between 50 and 100 resting hairs each day.

22. We are born with every hair follicle we'll ever have, though the composition, color, and pattern of the hair changes over time.

23. Hair is made of protein, so eating a healthy diet that contains sufficient protein, vitamins, minerals, and water is the best way to ensure healthy hair.

55 Phobias and Their Definitions

1. **Ablutophobia:** fear of washing or bathing

2. **Achluophobia:** fear of darkness

3. **Acrophobia:** fear of heights

4. **Aerophobia (or aviophobia):** fear of flying

5. **Agoraphobia:** fear of open spaces, crowds, or leaving a safe place

6. **Aichmophobia:** fear of pointed objects

7. **Ailurophobia:** fear of cats

8. **Alektorophobia:** fear of chickens

9. **Algophobia:** fear of pain

10. **Amaxophobia:** fear of riding in a car

11. **Anthropophobia:** fear of people or society

12. **Anuptaphobia:** fear of staying single

13. **Arachibutyrophobia:** fear of peanut butter sticking to the roof of your mouth

14. **Arachnophobia:** fear of spiders

15. **Astraphobia:** fear of thunder and lightning

16. **Atychiphobia:** fear of failure

17. **Autophobia:** fear of being alone

18. **Barophobia:** fear of gravity

19. **Bathmophobia:** fear of stairs or steep slopes

20. **Bibliophobia:** fear of books

21. **Catoptrophobia:** fear of mirrors

22. **Chronomentrophobia:** fear of clocks

23. **Chronophobia:** fear of time or the passage of time

24. **Claustrophobia:** fear of closed spaces

25. **Coulrophobia:** fear of clowns

26. **Cynophobia:** fear of dogs

27. **Dystychiphobia:** fear of accidents

28. **Emetophobia:** fear of vomiting

29. **Gamophobia:** fear of marriage

30. **Genuphobia:** fear of knees

31. **Glossophobia:** fear of public speaking

32. **Hemophobia:** fear of blood

33. **Hippopotomonstrosesquipedaliophobia:** fear of long words...go figure!

34. **Iatrophobia:** fear of doctors

35. **Ichthyophobia:** fear of fish

36. **Koumpounophobia:** fear of clothing buttons

37. **Lockiophobia:** fear of childbirth

38. **Melanophobia:** fear of the color black

39. **Mageirocophobia:** fear of cooking

40. **Mysophobia:** fear of germs or dirt

41. **Necrophobia:** fear of death or dead things

42. **Nomophobia:** fear of being without your mobile phone

43. **Nosocomephobia:** fear of hospitals

44. **Nyctophobia:** fear of the dark or of night

45. **Ophidiophobia:** fear of snakes

46. **Ornithophobia:** fear of birds

47. **Philemaphobia:** fear of kissing

48. **Philophobia:** fear of being in love

49. **Pupaphobia:** fear of puppets

50. **Pyrophobia:** fear of fire

51. **Scoptophobia:** fear of being stared at

52. **Trichophobia:** fear of hair

53. **Trypophobia:** fear of small holes or bumps

54. **Xanthophobia:** fear of the color yellow

55. **Zuigerphobia:** fear of vacuum cleaners

44 Facts About Sleep

1. By 60 years of age, 60 percent of men and 40 percent of women will snore.

2. While snores average around 60 decibels, the noise level of normal speech, they can reach more than 80 decibels.

3. Eighty decibels is as loud as the sound of a pneumatic drill breaking up concrete, and noise levels over 85 decibels are considered hazardous to the human ear.

4. Snoring can be a sign of sleep apnea, a life-threatening sleep disorder.

5. As many as 10 percent of people who snore have sleep apnea.

6. Sleep apnea can cause people to stop breathing as many as 300 times every night and can lead to a stroke or heart attack.

7. The longest time a human being has gone without sleep is 11 days and 25 minutes.

8. The world record was set by American 17-year-old Randy Gardner in 1963.

9. On average, humans sleep three hours less than other primates.

10. Chimps, rhesus monkeys, and baboons sleep ten hours per night.

11. When we sleep, we drift between rapid-eye-movement (REM) sleep and non-REM sleep in alternating 90-minute cycles.

12. Non-REM sleep starts with drowsiness and proceeds to deeper sleep, during which it's harder to be awakened.

13. During REM sleep, our heart rates increase, our breathing becomes irregular, our muscles relax, and our eyes move rapidly beneath our eyelids.

14. REM sleep was initially discovered years before the first studies that monitored brain waves overnight were conducted in 1953, though scientists didn't understand its significance at first.

15. Studies have shown that our bodies experience diminished capacity after we've been awake for just 17 hours.

16. We behave as if we were legally drunk.

17. After five nights with too little sleep, we actually get intoxicated twice as fast.

18. During their first year, babies cause between 400 and 750 hours of lost sleep for parents.

19. A newborn (0–3 months) needs 16.5 hours of sleep a day.

20. Infants (4–11 months) need about 12 to 15 hours of sleep per day.

21. Toddlers (1–2 years) should get 11 to 14 hours of sleep a day, while preschoolers (3–5 years) should average 10 to 13 hours a day.

22. The average adult needs 7 to 9 hours of sleep per day.

23. Some adults called "short sleepers" naturally require less than 6 hours of sleep at night.

24. At the other extreme are "long sleepers" who need 9 or more hours of sleep per night.

25. There is no evidence that seniors need less sleep than younger adults.

26. If you average 8 hours of sleep a night, you'll have slept for more than 100,000 hours by the time you're 35.

27. During sleep, the body manufactures a hormone that prevents movement, which is why you don't actually act out your dreams.

28. At least 10 percent of all people sleepwalk at least once in their lives.

29. Men are more likely to sleepwalk than women.

30. Sleepwalking occurs most commonly in middle childhood and preadolescence (11 to 12 years of age), and it often lasts into adulthood.

31. When we sleep, our bodies cool down. Body temperature and sleep are closely related.

32. Most people sleep best in moderate temperatures.

33. Caffeine can overcome drowsiness, but it actually takes about 30 minutes before its effects kick in, and they are only temporary.

34. Our eventual need for sleep is due in part to two substances the body produces—adenosine and melatonin.

35. While we're awake, the level of adenosine in the body continues to rise, signaling a shift toward sleep. While we sleep, the body breaks down adenosine.

36. When it gets dark, the body releases the hormone melatonin, which prepares the brain and body for sleep and helps us feel drowsy.

37. Most people average about four dreams per night, each lasting roughly 20 minutes.

38. People reaching 80 years old will have had approximately 131,400 dreams in their lifetimes.

39. Most people take about 6 seconds to yawn.

40. If you see someone yawn, there's a 55 percent chance you'll also yawn within 5 minutes.

41. There's a 65 percent chance you'll start yawning soon, just because you've been reading about yawning!

42. A study from 2003 found that 18 percent of people surveyed never or rarely dreamed in color.

43. In 1942 as many as 71 percent never or rarely dreamed in color.

44. Brain waves are actually more active during dreams than they are when a person is awake.

13 Elizabeth Blackwell Facts

1. Elizabeth Blackwell was the first woman to receive a medical degree in the United States.

2. She was born in 1821 in England. When she was a teenager, her family moved to New York and later to Ohio.

3. Blackwell was a schoolteacher and an active social reformer who believed in the abolition of slavery and women's rights.

4. Blackwell applied to and was rejected by several medical schools.

5. She was finally accepted in 1847 to Geneva Medical College in New York after the current crop of medical students voted to accept her.

6. The requirement was that the vote had to be unanimous: If even one male student had voted against her, Blackwell would have been denied.

7. Her graduate thesis discussed typhus fever.

8. She received her medical degree in 1849.

9. While working with an infant patient with an eye infection, she herself became infected and lost her sight in one eye.

10. She and her sister Emily, along with another female doctor, set up the New York Infirmary for Indigent Women and Children in 1853.

11. Today, that facility is the New York-Presbyterian Lower Manhattan Hospital.

12. Rebecca Cole, the second Black woman in the United States to become a physician, interned there.

13. Blackwell mentored Elizabeth Garrett Anderson, the first woman to become a physician in Britain.

Chapter 5

FOOD AND DRINK

42 Cheesy Facts

1. Cheddar cheese production involves "cheddaring"— the repeated cutting and piling of curds to create a firm cheese.

2. During World War II and for several years after, nearly all British cheese production was devoted to cheddar.

3. Fresh cheddar curds—the natural shape of the cheese before it's pressed into a block and aged—are called "squeaky curds."

4. Cheddar cheese is naturally white or pale yellow.

5. People started dying cheese orange back in the 17th century to prevent seasonal color variations.

6. These days, much of it is dyed orange with seeds from the annatto plant.

7. Early cheese makers used carrot juice and marigold petals.

8. Before 1850, nearly all cheese produced in the United States was cheddar.

9. Traditional English cheddar is produced in wheels and aged in cloth for a minimum of six months.

10. Someone who sells cheese professionally at a cheese shop or specialty food store is called a cheesemonger.

11. Age is the only thing that makes mild cheddar taste different from sharp cheddar.

12. Mild cheddar is usually aged for a few months.

13. The sharpest cheddars might be aged two years or more.

14. Archaeological surveys show that cheese was being made from the milk of cows and goats in Mesopotamia before 6000 B.C.

15. Travelers from Asia are thought to have brought the art of cheese making to Europe, where the process was adapted and improved in European monasteries.

16. American cheese is a kind of cheddar that undergoes additional processing.

17. Processed American cheese was developed in 1915 by J. L. Kraft as an alternative to traditional cheeses that had a short shelf life.

18. A one-ounce serving of cheese is about the size of four dice.

19. The Pilgrims brought cheese onboard the *Mayflower* in 1620.

20. Philadelphia cream cheese is named after a village in upstate New York, not the Pennsylvania city.

21. Cheese contains trace amounts of naturally occurring morphine, which comes from the cow's liver.

22. The holes in Swiss cheese were once seen as a sign of imperfection and something cheesemakers tried to avoid.

23. Cheddar has been around since at least the 1100s.

24. A purchase of more than 10,000 pounds of cheddar is listed in the financial records of King Henry II, dated 1170.

25. Cheddar is the most popular kind of cheese in world.

26. Legend has it that cheddar was first created when a milkmaid forgot a pail of milk in England's Cheddar Gorge caves.

27. By the time she returned, the milk had turned to cheese.

28. Queen Victoria received a 1,000-pound wheel of cheddar as a wedding present in 1840.

29. Wisconsin produces about 2.6 billion pounds of cheese each year.

30. A cheddar wheel displayed at the Toronto Industrial Exposition in the 1800s inspired the poem "Ode on the Mammoth Cheese Weighing over 7,000 Pounds" by James McIntyre.

31. British explorer Robert Falcon Scott brought 3,500 pounds of cheddar cheese with him on his ill-fated expedition to the South Pole in 1901.

32. Founded in 1882, the Crowley Cheese Factory in Healdville, Vermont, is the nation's oldest cheesemaker still in operation.

33. In 1964, the Wisconsin Cheese Foundation created a 34,665-pound wheel of cheddar to display at the 1964 World's Fair in New York City.

34. President Andrew Johnson once served a 1,400-pound block of cheddar at a White House party.

35. Wisconsin-based Simon's Specialty Cheese made a 40,060-pound cheddar wheel named the "Belle of Wisconsin" in 1988.

36. June is National Dairy Month, and the last week in June is National Cheese Week.

37. The Cheese Days celebration in Monroe, Wisconsin, has been held every other year since 1914.

38. Highlights include a 400-pound wheel of Swiss cheese and the world's largest cheese fondue.

39. The crunchy bits you sometimes get inside of aged cheese are amino acid crystals.

40. A cheesemaker in Quebec created a 57,518-pound cheddar cheese in 1995, claiming it to be the world's largest.

41. In 2010, chef Tanys Pullin sculpted a 1,100-pound crown out of cheddar cheese in honor of Queen Elizabeth II's coronation anniversary. It was the largest cheese sculpture in history at the time.

42. Stilton blue cheese frequently causes odd, vivid dreams.

10 Smelliest Cheeses

1. Limburger

2. Epoisses

3. Stinking Bishop

4. Stilton

5. Taleggio

6. Roquefort

7. Munster d'Alsace

8. Camembert

9. Pont l'Eveque

10. Serra da Estrela

28 Pretzel Facts

1. According to one origin story, Italian monks invented soft pretzels in A.D. 610 to motivate catechism students.

2. The monks called the treat *pretiola*, or "little rewards."

3. The twists resemble arms folded in prayer, and the three holes might represent the Father, Son, and Holy Spirit.

4. Pretzels are made of a simple mixture of water, flour, and salt.

5. In Germany, there are stories that pretzels were the invention of desperate bakers held hostage by local dignitaries.

6. Bakers in Austria have their own coat of arms: two lions holding a pretzel.

7. It was granted in 1510, after monks baking pretzels in a basement heard invaders digging tunnels under Vienna's city walls and helped defeat the invasion.

8. Pretzels show up in medieval religious art, including depictions of the Last Supper and in a prayer book created in 1440.

9. In 2015, archaeologists found a 250-year-old pretzel in Bavaria, perhaps the oldest ever discovered in Europe.

10. Hard pretzels were invented in the 1600s in Pennsylvania.

11. Legend has it that a baker's apprentice overcooked a batch of pretzels, accidentally creating this tasty snack.

12. In 1861, Julius Sturgis created the first commercial pretzel bakery in Lititz, Pennsylvania.

13. The Bavarian pretzel, or *Bayerische breze*, is on the European Union's protected origins list.

14. Only pretzels made in the state of Bavaria can be sold as Bavarian pretzels in the E.U.

15. The average American eats about two pounds of pretzels in a year.

16. Philadelphians eat about six times that amount.

17. Pretzels are a symbol of good luck and have been used in New Year's celebrations in Germany.

18. This food is also a symbol of eternal love. They've even been included in wedding ceremonies—the possible origin of the phrase "tying the knot."

19. The high school in Freeport, Illinois, nicknamed "Pretzel City USA" has a pretzel as a mascot.

20. The world's largest pretzel was baked in El Salvador in 2015.

21. It measured 29 feet 3 inches long by 13 feet 3 inches wide. It weighed 1,728 pounds.

22. President George W. Bush choked on a pretzel while watching a football game in 2002 and temporarily lost consciousness.

23. Americans buy more than $550 million worth of pretzels each year. About 80 percent are made in Pennsylvania.

24. In the United States, April 26 is National Pretzel Day.

25. Every pretzel in the world was made by hand until 1935, when the first automated pretzel machine was introduced in Pennsylvania.

26. Unsalted pretzels are nicknamed "baldies."

27. Pretzels have long been a beloved bar snack.

28. In fact, Prohibition in the 1920s hit the pretzel business hard as bars were legally required to close their doors.

44 Amazing Alcohol Facts

1. The name whiskey is the result of a mispronunciation of the Gaelic *uisge beatha*, which literally means "water of life" and was commonly used to refer to distilled liquor.

2. The oldest evidence of alcoholic beverages ever found dates as far back as 7000 B.C. in China.

3. The workers who built Egypt's Great Pyramids may have received their payment in beer.

4. It takes about six minutes for alcohol to start affecting the brain.

5. The Code of Hammurabi, ancient Babylon's set of laws, stipulated a daily ration of beer for citizens.

6. Gin and tonics originated as a way to make anti-malarial medication more palatable.

7. Sloe gin isn't gin at all. It's a liqueur made with sloe berries (blackthorn bush berries).

8. About 600 grapes go into one bottle of red wine.

9. As they age, white wine gets darker and red wine gets lighter.

10. It can take 40 years for vintage Port to reach full maturity.

11. Every bottle of Champagne has an estimated 49 million bubbles.

12. The world's oldest existing bottle of wine was buried in A.D. 350 in Germany.

13. It was found again in 1867 and is now on display at a museum.

14. California produces more wine than any country in the world, with the exception of France, Italy, and Spain.

15. Kentucky produces all but five percent of the world's bourbon.

16. Tequila has no worm in it.

17. Mezcal does, although it's not even a worm, it's a moth larva.

18. Despite the perceived warming effects, alcohol actually lowers your body temperature.

19. Early thermometers often contained brandy instead of mercury.

20. A particularly drunken Christmas party at West Point Academy in 1826 resulted in a riot and 19 people being expelled.

21. The incident became known as the Eggnog Riot.

22. During Prohibition in the United States, doctors could still prescribe whiskey to patients.

23. This loophole helped Walgreens grow exponentially in those years.

24. Prohibition in the U.S. created a boom for Canadian distillers, and rye (the Canadian name for Canadian whiskey) became part of the national identity.

25. Rye is generally sweeter than bourbon and retains a worldwide following among liquor connoisseurs.

26. The exact origin of absinthe is unknown, but the strong alcoholic liqueur was probably first commercially produced around 1797.

27. It takes its name from one of its ingredients, *Artemisia absinthium*, which is the botanical name for the bitter herb known as wormwood.

28. Green in color due to the presence of chlorophyll, absinthe became an immensely popular drink in France by the 1850s.

29. The United States banned absinthe in 1912, years before Prohibition, and it remained banned until 2007.

30. Most Americans consider sake a Japanese rice wine, but it's actually more akin to beer.

31. Sake also may have originated in China, not Japan.

32. The 1808 Rum Rebellion in Australia kicked out New South Wales governor William Bligh when he got in the way of the colony's rum business, among other things.

33. Contrary to urban legend, Jägermeister does not contain elk blood.

34. The founder was an avid hunter, and the name Jägermeister literally translates to "hunt master."

35. What's the difference between brandy and cognac? All cognac is brandy, but not all brandy is cognac.

36. Brandy refers to any distilled spirit made from fermented fruit juice.

37. Cognac is a type of brandy from the Cognac region of France.

38. Strict rules govern where the grapes must be grown, and how the spirit is then produced.

39. Tarantula brandy is a popular drink in Cambodia.

40. The main difference between scotch and whiskey is geographic. Ingredients and spellings can differ, too.

41. Scotch is whisky made in Scotland (and spelled without an "e"), while bourbon is whiskey made in the U.S.

42. Scotch is made mostly from malted barley, while bourbon is distilled from corn.

43. If you ask for a whisky in England, you'll get scotch. But in Ireland, you'll get Irish whiskey (they spell it with an "e").

44. After a Tennessee whiskey, like Jack Daniel's, for example, is distilled, it's filtered through sugar-maple charcoal.

31 Mush-ruminations

1. France was the first country to cultivate mushrooms, in the mid-17th century.

2. From there, the practice spread to England and made its way to the United States in the 19th century.

3. In 1891, New Yorker William Falconer published *Mushrooms: How to Grow Them—A Practical Treatise on Mushroom Culture for Profit and Pleasure*, the first book on the subject.

4. In North America alone, there are an estimated 10,000 species of mushrooms, only 250 of which are known to be edible.

5. A mushroom is a fungus (from the Greek word *sphongos*, meaning "sponge").

6. A fungus differs from a plant in that it has no chlorophyll, produces spores instead of seeds, and survives by feeding off other organic matter.

7. Mushrooms are related to yeast, mold, and mildew, which are also members of the "fungus" class.

8. There are approximately 1.5 million species of fungi, compared with 250,000 species of flowering plants.

9. An expert in mushrooms and other fungi is called a mycologist—from the Greek word *mykes*, meaning "fungus."

10. A mycophile is a person whose hobby is to hunt edible wild mushrooms.

11. Ancient Egyptians believed mushrooms were the plant of immortality.

12. Pharaohs decreed them a royal food and forbade commoners to even touch them.

13. White agaricus (aka "button") mushrooms are by far the most popular, accounting for more than 90 percent of mushrooms bought in the United States each year.

14. Brown agaricus mushrooms include cremini and portobellos, though they're really the same thing: Portobellos are just mature cremini.

15. Cultivated mushrooms are agaricus mushrooms grown on farms.

16. Exotics are any farmed mushroom other than agaricus (think shiitake, maitake, oyster).

17. Wild mushrooms are harvested wherever they grow naturally—in forests, near riverbanks, even in your backyard.

18. Many edible mushrooms have poisonous look-alikes in the wild.

19. For example, the dangerous "yellow stainer" closely resembles the popular white agaricus mushroom.

20. "Toadstool" is the term often used to refer to poisonous fungi.

21. In the wild, mushroom spores are spread by wind.

22. On mushroom farms, spores are collected in a laboratory and then used to inoculate grains to create "spawn," a mushroom farmer's equivalent of seeds.

23. A mature mushroom will drop as many as 16 billion spores.

24. Mushroom spores are so tiny that 2,500 arranged end-to-end would measure only an inch in length.

25. Mushroom farmers plant the spawn in trays of pasteurized compost, a growing medium consisting of straw, corn-cobs, nitrogen supplements, and other organic matter.

26. The process of cultivating mushrooms—from preparing the compost in which they grow to shipping the crop to markets—takes about four months.

27. The small town of Kennett Square, Pennsylvania, calls itself the Mushroom Capital of the World—producing more than 51 percent of the nation's supply.

28. September is National Mushroom Month.

29. One serving of button mushrooms (about five) has only 20 calories and no fat.

30. Mushrooms provide such key nutrients as B vitamins, copper, selenium, and potassium.

31. Some experts say the taste of mushrooms belongs to a "fifth flavor"—beyond sweet, sour, salty, and bitter—known as umami, from the Japanese word meaning "delicious."

38 Revolting Food Facts

1. Baby mice wine, which is made by preserving newborn mice in a bottle of rice wine, is a traditional health tonic from Korea said to aid the rejuvenation of one's vital organs.

2. Ever get a hankering for soft-boiled duck embryos? Balut are duck eggs that have been incubated for 15 to 20 days (a duckling takes 28 days to hatch) and then boiled.

3. The egg is then consumed—both the runny yolk and the beaky, feathery, veiny duck fetus. Balut are eaten in the Philippines, Cambodia, and Vietnam.

4. Black pudding, eaten in Britain and Ireland, is congealed pig blood that's been cooked with oatmeal and formed into a small disk. It tastes like a thick, rich, beef pound cake.

5. The Sardinian delicacy casu marzu is a hard sheep's milk cheese infested with *Piophila casei*, the "cheese fly."

6. The larvae produce enzymes that break down the cheese into a tangy goo, which Sardinians dive into and enjoy, larvae and all.

7. Surströmming, primarily a seasonal dish in northern Sweden, is rotten fermented herring. Even the Swedes rarely open a can of it indoors.

8. History shows that people will make alcohol from any ingredients available. That includes bananas, mashed with one's bare feet and buried in a cask. The result is pombe, an east African form of beer.

9. Cobra heart delivers precisely what it promises: a beating cobra heart, sometimes accompanied by a cobra kidney and chased by a slug of cobra blood.

10. Preparations of the Vietnamese delicacy involve a large blade and a live cobra.

11. Escamoles are the eggs, or larvae, of the giant venomous black *Liometopum* ant.

12. This savory Mexican chow has the consistency of cottage cheese and a surprisingly buttery and nutty flavor.

13. The key ingredient of bird's nest soup is the saliva-rich nest of the cave swiftlet, a swallow that lives on cave walls in Southeast Asia.

14. Hákarl, an Icelandic dish dating back to the Vikings, is putrefied shark meat.

15. Traditionally, it has been prepared by burying a side of shark in gravel for three months or more.

16. Nowadays, hákarl might be boiled in several changes of water or soaked in a large vat filled with brine and then cured in the open air for two months.

17. This is done to purge the shark meat of urine and trimethylamine oxide.

18. Durian is a football-size fruit with spines.

19. It smells like unwashed socks but tastes sweet. Imagine eating vanilla pudding while trying not to inhale.

20. Lutefisk is a traditional Scandinavian dish made by steeping pieces of cod in lye solution.

21. The result is translucent and gelatinous, stinks to high heaven, and corrodes metal kitchenware.

22. Enjoy lutefisk covered with pork drippings, white sauce, or melted butter, with potatoes and Norwegian flatbread on the side.

23. Kava is the social lubricant of many island nations. Take a pepper shrub root (*piper methysticum*), and get someone to chew or grind it into a pulp. Mix with water and enjoy.

24. Pacha is a sheep's head stewed, boiled, or otherwise slow-cooked for five to six hours together with the sheep's intestines, stomach, and feet. Yum!

25. Vegemite, made from leftover brewers' yeast extract, looks like chocolate spread, smells like B vitamins, and tastes overwhelmingly salty.

26. Australians love Vegemite on sandwiches or baked in meatloaf.

27. When you're a Mongolian nomad and there are no taverns, you're happy to settle for fermented mare's milk, airag.

28. It takes only a couple of days to ferment and turns out lightly carbonated.

29. Spiders are popular fare in parts of Cambodia, especially in the town of Skuon.

30. Tarantulas are sold on the streets and are said to be very good fried with salt, pepper, and garlic.

31. If you find yourself hungry as you hustle through London, grab a jellied eel from a street vendor.

32. It tastes like pickled herring with a note of vinegar, salt, and pimiento, all packed in gelatin.

33. Drink enough Scotch, and you'll eventually get so hungry you'll eat haggis—sheep innards mixed with oatmeal and boiled in the sheep's stomach.

34. Some tribespeople in Papua New Guinea consider Sago beetle grubs delicious.

35. A specialty of Newfoundland, seal flipper pie is made from the chewy cartilage-rich flippers of seals, usually cooked in fatback with root vegetables and sealed in a flaky pastry crust or topped with dumplings.

36. Eating improperly prepared pufferfish can result in sudden death. As such, Japan requires extensive training

and apprenticeship, as well as special licensing, for the chefs preparing this highly sought-after delicacy.

37. Menudo is basically cow-stomach soup.

38. If you can tolerate the slimy, rubbery tripe chunks, the soup itself tastes fine. It's often served in Mexico for breakfast to cure a hangover.

51 Popcorn Facts

1. Popcorn's scientific name is *zea mays everta*.

2. It is the only type of corn that will pop.

3. People have been enjoying popcorn for thousands of years.

4. The first evidence of popcorn has been radiocarbon-dated as 6,700 years old (c. 4700 B.C.), based on macrofossil cobs unearthed between 2007 and 2011 at archaeological sites on the northern coast of Peru.

5. It is believed that the Wampanoag American Indian tribe brought popcorn to the Pilgrims for the first Thanksgiving in Plymouth, Massachusetts.

6. Traditionally, American Indian tribes flavored popcorn with dried herbs and spices, possibly even chili.

7. They also made popcorn into soup and beer, and made popcorn headdresses and corsages.

8. Some American Indian tribes believed that a spirit lived inside each kernel of popcorn.

9. The spirits wouldn't usually bother humans, but if their home was heated, they would jump around, getting angrier and angrier, until eventually they would burst out with a pop.

10. Christopher Columbus allegedly introduced popcorn to the Europeans in the late 15th century.

11. Charles Cretors invented the first commercial popcorn machine in Chicago in 1885.

12. The business he founded still manufactures popcorn machines and other specialty equipment.

13. American vendors began selling popcorn at carnivals in the late 19th century.

14. Movie theater owners initially feared that popcorn would distract their patrons.

15. It took a few years for them to realize that popcorn could be a way to increase revenues, and popcorn has been served in movie theaters since 1912.

16. Nowadays, many movie theaters make a greater profit from popcorn than they do from ticket sales.

17. Popcorn also makes moviegoers thirsty and more likely to buy expensive drinks.

18. Each popcorn kernel contains a small amount of moisture.

19. As the kernel is heated, this water turns to steam.

20. The popcorn kernel's shell is not water-permeable, so the steam cannot escape and pressure builds up until the kernel finally explodes, turning inside out.

21. On average, a kernel will pop when it reaches a temperature of 347 °F (175 °C).

22. Unpopped kernels are called "old maids" or "spinsters."

23. There are two possible explanations for old maids.

24. The first is that they didn't contain sufficient moisture to create an explosion.

25. The second is that their outer coating (the hull) was damaged, so that steam escaped gradually, rather than with a pop.

26. Good popcorn should produce less than two percent old maids.

27. Ideally, the moisture content of popcorn should be around 13.5 percent, as this results in the fewest old maids.

28. Popcorn is naturally high in fiber, low in calories, and sodium-, sugar-, and fat-free, although oil is often added during preparation, and butter, sugar, and salt are popular toppings.

29. Americans consume 17 billion quarts of popped popcorn each year.

30. That's enough to fill the Empire State Building 18 times!

31. Nebraska produces more popcorn than any other state in the country—around 250 million pounds per year.

32. That's about a quarter of all the popcorn produced annually in the United States.

33. There are at least five contenders claiming to be the "Popcorn Capital of the World" due to the importance of popcorn to their local economies: Van Buren, Indiana; Marion, Ohio; Ridgway, Illinois; Schaller, Iowa; and North Loup, Nebraska.

34. Popped popcorn comes in two basic shapes: snowflake and mushroom.

35. Movie theaters prefer snowflake because it's bigger.

36. Confections such as caramel corn use mushroom because it won't crumble.

37. The employees at The Popcorn Factory in Lake Forest, Illinois, made a 3,423-pound popcorn ball in 2006.

38. It was the world's largest popcorn ball at the time.

39. You can visit the World's Largest Popcorn Ball in Sac City, Iowa.

40. Built in 2016, the popcorn ball weighs 9,370 pounds and stands taller than eight feet.

41. In 1948, ears of popcorn (the variety of corn grown for this particular purpose) were discovered in the Bat Cave in New Mexico.

42. They were around 5,600 years old.

43. Aztec Indians of the 16th century used garlands of popped maize as a decoration in ceremonial dances.

44. They were a symbol of goodwill and peace.

45. Colonial women poured sugar and cream on popcorn and served it for breakfast—likely the first "puffed" cereal!

46. Some colonists popped corn in a type of cage that revolved on an axle and was positioned over a fire.

47. During the Great Depression, popcorn was ubiquitous on city streets because it was an inexpensive way to stave off hunger.

48. According to the Popcorn Institute, popcorn is high in carbohydrates and has more protein and iron than potato chips, pretzels, and soda crackers.

49. If you made a trail of popcorn from New York City to Los Angeles, you would need more than 352,028,160 popped kernels!

50. The world's longest popcorn string measured 1,200 feet.

51. It was created by Gary Kohs, Laura Scaccia, and Dave Vandenbossche at Drake Park in Marine City, Michigan, in October 2016.

34 Fortune Cookie Facts

1. To most Americans, fortune cookies are synonymous with Chinese food, but they didn't originate in China.

2. It's extremely difficult to find fortune cookies in China.

3. This is because the fortune cookie actually traces its origins back to Japan, not China.

4. Fortune cookies are a mainstay in the United States, but they are also served in Britain, Italy, France, and Mexico.

5. More than three billion fortune cookies are made each year, the vast majority of them in the United States.

6. The recipe for fortune cookies is surprisingly simple. All you need is flour, sugar, vanilla, and sesame oil.

7. Most manufacturers add other ingredients. Some use vegetable shortening or butter instead of sesame oil. Starch, eggs, and food coloring are also regularly used.

8. No one knows for sure who first introduced the fortune cookie to the United States, but two entrepreneurs in California are given credit.

9. According to one legend, Japanese immigrant Makoto Hagiwara, a landscape designer responsible for Golden Gate Park's Japanese Tea Garden, introduced the first U.S. fortune cookie in 1914 in San Francisco.

10. A second legend credits Chinese immigrant David Jung, the founder of a noodle company, with introducing the cookie in 1918 in Los Angeles.

11. According to the story, Jung was concerned about the number of poor people living on the streets, so he passed out free fortune cookies to them.

12. Each cookie contained an inspirational verse written for Jung by a Presbyterian minister.

13. In 1983, San Francisco's Court of Historical Review held a mock trial to determine whether Hagiwara or Jung should get credit for bringing fortune cookies to U.S. diners.

14. Not surprisingly, the judge ruled for San Francisco and Hagiwara.

15. A piece of evidence that surfaced during the trial was a fortune saying, "S.F. judge who rules for L.A. not very smart cookie."

16. As far back as the 19th century, a cookie very similar in appearance to the modern fortune cookie was made in Kyoto, Japan.

17. A woodblock image from 1878 shows what seems to be a Japanese street vendor grilling fortune cookies.

18. The Japanese version of the cookie is a bit larger and darker than the American fortune cookie.

19. Their batter contains miso paste and sesame rather than vanilla and butter.

20. The fortunes were never put inside the cookies either.

21. Instead, they were tucked into the fold of the fortune cookie on the outside.

22. This kind of cookie is called *tsujiura senbei*.

23. Fortune cookies became common in Chinese restaurants after World War II.

24. Edward Louie, owner of the Lotus Fortune Cookie Company in San Francisco, invented a machine in 1974 that could insert the fortune and fold the cookie.

25. In 1980, Yong Lee created an upgraded, fully-automated fortune-cookie machine.

26. Wonton Food, Inc., in New York, is the largest producer of fortune cookies in the United States.

27. The factory churns out 4.5 million cookies per day.

28. The company boasts a database of 15,000 possible fortunes.

29. Company officials say that only about 25 percent of these fortunes are used at any given time.

30. Some fortune cookies don't contain fortunes at all. Crack one open, and you'll often find lucky lottery numbers or a philosophical message.

31. Fortune cookies today come in a wide variety of flavors.

32. Some are covered in chocolate or caramel. Many bakeries also sell fortune cookies decorated for Christmas, Valentine's Day, and other holidays.

33. Unless you really love them, you won't gain too much weight eating fortune cookies.

34. The average fortune cookie contains about 30 calories and no fat.

29 Wacky Jelly Belly Flavors

1. Pickle

2. Black Pepper

3. Booger

4. Dirt

5. Earthworm

6. Earwax

7. Sausage

8. Rotten Egg

9. Soap

10. Vomit

11. Sardine

12. Grass

13. Skunk Spray

14. Bacon

15. Baby Wipes

16. Pencil Shavings

17. Toothpaste

18. Moldy Cheese

19. Buttered Popcorn

20. Dr. Pepper

21. Jalapeño

22. Margarita

23. Spinach

24. Cappuccino

25. Peanut Butter

26. Café Latte

27. 7UP

28. Pomegranate

29. Baked Beans

13 Vodka Facts

1. This colorless alcohol hails from Russia, where its original name is *zhiznennaia voda*, or "water of life."

2. Vodka may be made with vegetables (such as potatoes or beets) or grains (barley, wheat, rye, or corn).

3. Grain vodka is considered higher quality.

4. Early vodkas were crudely manufactured and tasted pretty bad.

5. It was common to mix vodka with herbs, spices, or honey to mask the harsh, offensive taste.

6. Today's distillation processes create clean-tasting vodkas, many of which are enhanced with vanilla or fruit flavors.

7. In 1540, Ivan the Terrible stopped fighting long enough to establish the country's first vodka monopoly.

8. In the late 17th century, Peter the Great explored improved methods of distillation and means of export.

9. During the reign of Peter the Great, it was customary that foreign ambassadors visiting Russia consume a liter and a half of vodka.

10. Lightweight ambassadors began to enlist stand-ins to drink so that the official could discuss important matters with a clear head.

11. Since vodka is an alcohol, consider using it to clean razors—the liquid disinfects the blade and prevents rusting.

12. Vodka can help remove bandages.

13. Saturating the bandage with vodka will dissolve the adhesive, making removal of the bandage painless.

34 Chocolaty Facts

1. Archaeologists have found cocoa residue in pottery unearthed in Honduras from 1400 B.C.

2. Four hundred cocoa beans go into each pound of chocolate.

3. A cacao tree is between four and five years old before it grows its first beans.

4. It takes about a year for a single cacao tree to produce enough beans to make 10 Hershey bars.

5. Cacao trees can live for hundreds of years, but they produce quality cocoa beans for only 25 years.

6. The Maya used cocoa beans as currency.

7. Chocolate helped researchers decode the Maya's written language.

8. The word *ka-ka-w*, or cacao, was written on containers of chocolate found buried with Maya dignitaries.

9. The cacao tree's scientific name is *Theobroma cacao*, or "food of the gods."

10. The cacao tree is in the same family as okra and cotton.

11. Cacao trees are native to Central and South America, but most cacao now comes from West Africa, primarily Côte d'Ivoire.

12. Chocolate was only enjoyed as a beverage for centuries.

13. The Aztec mixed cacao seeds with chilies to make a frothy, spicy drink.

14. Aztec Emperor Montezuma is said to have drunk 50 goblets of chocolate, flavored with chili peppers, every day.

15. Explorer Hernán Cortés brought the Aztec drink back to Spain.

16. French and Spanish nobles added cinnamon and cane sugar to sweeten the bitter beverage.

17. In 1847, Joseph Fry & Sons in England made the first "eating" chocolate bar.

18. The ancient Olmec of present-day Mexico and Central America are believed to be the first people to use the cacao plant, as early as 1000 B.C.

19. A single fruit of the cacao tree contains between 20 and 60 cocoa seeds.

20. The "blood" running down the drain in *Psycho*'s shower scene was actually chocolate syrup.

21. If you're feeling down after watching a sad film, eating chocolate can improve your mood.

22. German chocolate cake isn't German.

23. It was named after an American named Sam German.

24. The familiar Hershey's chocolate bar was first sold in 1900.

25. The first Hershey's Kisses were introduced in 1907.

26. White chocolate technically isn't chocolate.

27. It's made with only cocoa butter, not cocoa powder, as real chocolate bars are.

28. People in Switzerland consume more chocolate than anyone else in the world, with an average resident going through nearly 20 pounds each year.

29. Chocolate-covered bacon is a popular snack at state fairs.

30. In 2007, a man used chocolate to help him make friends with some guards at a Belgian bank.

31. After gaining the guards' trust, the man calmly walked out of the bank with $28 million worth of stolen diamonds.

32. A study published in 2016 found people were more likely to make purchases if a store smelled like chocolate.

33. Some researchers have found that people rate the quality of a chocolate higher if they eat it while looking at a nice painting.

34. The Baby Ruth candy bar was not named for baseball slugger Babe Ruth but actually for President Grover Cleveland's daughter Ruth.

10 Food Origins

1. **Pizza:** Pizza originated in the South Italian region around Naples, where flat yeast-based bread topped with tomatoes was a local specialty. In 1889, baker Raffaele Esposito made pizzas for the visiting Italian King Umberto I and Queen Margherita. The queen's favorite pizza featured basil leaves over mozzarella cheese and tomatoes—a dish soon known as a Margherita pizza.

2. **Sandwich:** This lunchtime favorite likely dates back to the ancient Hebrews, who may have put meat and herbs between unleavened bread during Passover. But an 18th-century British noble, John Montagu, 4th Earl of Sandwich, gave it its name. The earl was a keen card player and commonly ate meat between pieces of bread to keep from getting the cards greasy.

3. **Caesar Salad:** Caesar Salad isn't named after Julius or Augustus. It's the creation of Italian-born Mexican chef Caesar Cardini, who according to one story, whipped up the salad when faced with a shortage of ingredients for a Fourth of July celebration in 1924. Another tale attests that Cardini made it for a gourmet contest in Tijuana. Either way, the salad is worthy of his name!

4. **Ranch Dressing:** Ranch Dressing actually did start out on a ranch—a dude ranch in Santa Barbara, California. Opened in 1954 by Steve and Gayle Henson, the "Hidden Valley Ranch" served a special house dressing that was so popular visitors came just to buy it.

5. **Taco:** The taco is one of the first "fusion" foods—as much a mix of cultures as the Mexican people today. The native Nahuatl people ate fish served in flat corn bread, but it was 16th-century Spanish explorers who gave the bread the name *tortilla* and began filling it with beef and chicken as well.

6. **Steak Tartare:** The nomadic Tartar warrior tribe was known for eating raw meat, which was usually pressed underneath the saddle of a horse, but the raw meat dish today, which is usually chopped beef or horse-meat with a liberal amount of seasoning and spices, may take its name from the Italian word *tartari*, which means raw steak.

7. **Pasty:** The forerunner of the potpie was originally cooked up as a lunch meal by Cornish miners, who were unable to return to the surface to eat. The pastry crust allowed the miners the ability to eat the contents,

which typically included meat, vegetables, and gravy, and then discard the shell.

8. **Thousand Island Dressing:** It turns out thousand island dressing owes its name to the "Thousand Islands" region in upstate New York, which actually boasts 1,800 small islands. Back in the early 1900s, actress May Irwin was served this scrumptious salad dressing while at a dinner party in the Thousand Islands area. She named the dressing and spread the word.

9. **Waffles:** During the Middle Ages, a thin crisp cake was baked between wafer irons. Oftentimes, the irons included designs that helped advertise the kitchen that produced the waffle. Early waffles were made of barley and oats, but by the 18th century, the ingredients changed to the modern version of leaven flour.

10. **Frozen Dinners:** Swanson's TV Brand Frozen Dinner was a popular prime-time meal in 1953, so it may come as a surprise that frozen dinners predated Swanson's by nearly a decade. William L. Maxson devised a prepackaged frozen meal for airplanes in 1944. Still, clever marketing and better distribution ensured that the TV Dinner from Swanson became the nationally recognized leader in frozen food.

21 Canadian Foods

1. Arctic char is the northernmost freshwater fish in North America, caught commercially since the 1940s. Char is a little like salmon in color and texture, but its unique flavor elevates it to a delicacy.

2. Back bacon, also called Canadian bacon in the U.S., has less fat than other kinds of bacon. Its taste and texture are similar to ham. Peameal bacon is cured back bacon that's coated with ground yellow peas.

3. Bakeapples are also referred to as baked-apple berries, chicoute, and cloudberries. Found mostly in the prov-

inces of Nova Scotia and Newfoundland, they taste—surprise—like a baked apple. These can be eaten raw or used in pies and jams.

4. To ward off the cold, Newfoundlanders have long boiled up bangbelly—a pudding of flour, rice, raisins, pork, spices, molasses, and sometimes seal fat. The result is comparable to bread pudding and is commonly served at Christmastime.

5. No aquatic rodents are harmed in the making of beavertails, an Ottawa specialty. These fried, flat pastries, shaped like a beaver's tail, are similar to that carnival staple called elephant ears. Top with sugar, cinnamon, fruit, even cream cheese and salmon.

6. Similar to the Bloody Mary, the Bloody Caesar cocktail is popular all over Canada. It's made of vodka, tomato-clam juice, Worcestershire sauce, and hot sauce, served on the rocks in a glass rimmed with celery salt.

7. Along with so many other aspects of their culture, the Scots brought butter tart to Canada. These are little pecan pies without the pecans, perhaps with chocolate chips, raisins, or nuts. You haven't had Canadian cuisine until you've had a butter tart.

8. Cipaille, a layered, spiced-meat-and-potato pie, is most popular in Quebec. Look for it on menus as "sea pie" in Ontario, not for any aquatic additives but because that's exactly how the French word is pronounced in English.

9. Break your morning butter-and-jam routine and have some cretons instead! This Quebecois tradition is a seasoned pork-and-onion pâté often spread on toast.

10. Dulse is a tasty, nutritious, protein-packed seaweed that washes up on the shores of Atlantic Canada. It's used in cooking much the same ways one uses onions: chopped, sautéed, and added to everything from omelets to bread dough.

11. Figgy duff is a traditional bag pudding from Newfoundland. Common ingredients include bread-crumbs, raisins, brown sugar, molasses, butter, flour, and spices.

12. Jigg's dinner is a traditional meal from Newfoundland that incorporates salt beef, cabbage, boiled potatoes, carrots, turnips, and homemade pease pudding.

13. Many consider the Prince Edward Island delicacies called malpeques the world's tastiest oysters, harvested with great care by workers who rake them out of the mud by hand. If you can find them, the "pride of P.E.I." will cost you dearly.

14. The Nanaimo bar is a chocolate bar layered with nuts, buttercream, and sometimes peanut butter or coconut. Nanaimo bars are well liked throughout Canada and in bordering U.S. regions (especially around Seattle).

15. Close to 90 percent of Canada's maple syrup comes from Quebec, and Canada is the world's largest pro-ducer of this sweet, sticky pancake topping.

16. Moosehunters are what Canadians call molasses cookies.

17. Nougabricot is a Quebecois preserve consisting of apri-cots, almonds, and pistachios.

18. Canada's large waves of Slavic immigration have brought pierogi to the True North. They're small dump-lings with a variety of fillings, including cheese, meat, potatoes, mushrooms, cabbage, and more. Top with sour cream and onions.

19. Persians are sort of a cross between a large cinnamon bun and a doughnut, topped with strawberry icing; unique to Thunder Bay, Ontario.

20. Acadia, the Cajuns' ancestral homeland, loves its buck-wheat pancakes. But the greenish-yellow griddle cakes known as ployes contain no milk or eggs, so they're not

actually pancakes. Eat ployes with berries, whipped cream, cretons, or maple syrup.

21. Dump gravy and cheese curds on French fries: Voilà la poutine! Quebec is the homeland of poutine, but you can get it all over the nation.

40 Coffee Facts

1. Coffee is one of the most popular beverages in the world, with more than 400 billion cups consumed each year.

2. Coffee is brewed from roasted beans of the plant species *Coffea*.

3. Coffee beans are technically fruit pits, not beans.

4. Coffee is believed to originate in Ethiopia.

5. According to legend, a goat herder named Kaldi noticed his goats became energetic after eating the berries from a certain tree.

6. He reported his findings to the abbot of a local monastery, who made a drink with the berries.

7. Coffee cultivation and trade began on the Arabian Peninsula.

8. By the 16th century, coffee houses were common in Persia, Egypt, Syria, and Turkey.

9. Brazil is the top coffee producing country, accounting for roughly 40 percent of the world's coffee supply.

10. The second-largest coffee producer is Vietnam, responsible for about 20 percent of global supply.

11. Around 25 million farmers worldwide depend on coffee crops for their economic livelihood.

12. After oil, coffee is the world's most valuable traded commodity.

13. Coffee contains caffeine, the stimulant that increases your metabolism.

14. Caffeine is the most popular drug in the world, and 90 percent of people in the U.S. consume it in some form every day.

15. It takes a day to fully eliminate caffeine from your system.

16. The effects of caffeine reach its peak around 30 to 60 minutes after consumption.

17. There are two main types of coffee: Arabica and Robusta.

18. Dark-roast coffee actually has less caffeine than coffee that's been lightly roasted.

19. The word coffee comes from Kaffa, a region in Ethiopia where coffee beans may have been discovered.

20. As early as the ninth century, people in the Ethiopian highlands were making a stout drink from ground coffee beans boiled in water.

21. By the 17th century, coffee had made its way to Europe and was becoming popular across the continent.

22. Dutch merchants shipped live coffee plants from the Yemeni port of Mocha to India and Indonesia, where they were grown on plantations to supply beans to Europe.

23. Some Europeans feared the new beverage, calling it the "bitter invention of Satan."

24. Pope Clement VIII (1536–1605) gave coffee papal approval after trying a cup.

25. By the middle of the 17th century, there were over 300 coffee houses in London.

26. In 1674, the Women's Petition Against Coffee claimed the beverage was turning British men into "useless corpses" and proposed a ban on it for anyone under the age of 60.

27. Scandinavia boasts the highest per-capita coffee consumption in the world.

28. On average, people in Finland drink more than four cups of coffee a day.

29. The traditional Finnish way of brewing coffee is a variation on Turkish coffee where water and coffee grounds are brought just barely to a boil repeatedly.

30. Espresso contains three times more caffeine per ounce than standard brewed coffee.

31. The first webcam was set up to watch a coffeepot so computer scientists at Cambridge University could monitor the coffee situation without leaving their desks.

32. The world's most expensive coffee, Kopi Luwak, sells for $100 to $600 per pound.

33. The pricy coffee beans are collected from Indonesian civet droppings.

34. The civet (a small weasel-like critter) eats coffee cherries but can't digest the beans, so they pass through the animal's digestive tract and are handpicked by locals.

35. Germany banned coffee pods in all government buildings.

36. The first "instant coffee" appeared in Britain in 1771.

37. It was called a "coffee compound" and had a patent granted by the British government.

38. In 1971, a group of Seattle-based entrepreneurs opened a coffee shop called Starbucks.

39. Americans consume an average of 9.7 pounds of coffee per year, making the U.S. only the 25th biggest consumer of coffee worldwide on a per-person basis.

40. The average person in the U.S. consumes about three cups of coffee per day.

20 Top Per Capita Coffee Consumers

Country—Pounds per Person per Year

1. Finland—26.45 lbs.
2. Norway—21.82 lbs.
3. Iceland—19.84 lbs.
4. Denmark—19.18 lbs.
5. Netherlands—18.52 lbs.
6. Sweden—18 lbs.
7. Switzerland—17.42 lbs.
8. Belgium—15 lbs.
9. Luxembourg—14.33 lbs.
10. Canada—14.33 lbs.
11. Bosnia and Herzegovina—13.67 lbs.
12. Austria—13.45 lbs.
13. Italy—13 lbs.
14. Brazil (tied)—12.79 lbs.
15. Slovenia (tied)—12.79 lbs.
16. Germany—12.13 lbs.
17. Greece (tied)—11.9 lbs.
18. France (tied)—11.9 lbs.
19. Croatia—11.24 lbs.
20. Cyprus—10.8 lbs.

17 Apple Facts

1. Apples are a member of the rose family.
2. There are approximately 7,500 varieties of apples grown around the world.

3. The crabapple is the only apple native to North America.

4. Apples will ripen six times faster if you leave them at room temperature rather than refrigerate them.

5. A peck of apples weighs 10.5 pounds.

6. A bushel of apples weighs 42 pounds, and will yield 20 to 24 quarts of applesauce.

7. Apple trees produce fruit four to five years after they're planted.

8. The science of apple growing is called pomology.

9. It takes about 36 apples to create one gallon of apple cider.

10. About 25 percent of an apple's volume is air; that's why they float.

11. Around two pounds of apples are needed to make one 9-inch pie.

12. The largest apple picked weighed three pounds.

13. A medium apple is about 80 calories.

14. A large-sized apple has about 130 calories.

15. Early American apple orchards produced very few apples because there were no honeybees.

16. Historical records indicate that colonies of honeybees were first shipped from England and landed in the colony of Virginia early in 1622.

17. Archeologists have found evidence that humans have been enjoying apples since at least 6500 B.C.

48 Fast Food Facts

1. Consumption statistics show that Americans eat more fast food than any other country.

2. The United States is home to more than 200,000 fast food restaurants.

3. France ranks second in fast food consumption.

4. KFC's iconic founder, Colonel Sanders, was never in the military.

5. Kentucky Governor Ruby Laffoon named Harland Sanders an honorary colonel.

6. Kentucky Fried Chicken was the brainchild of Harland Sanders, who opened his first restaurant during the Great Depression in a gas station in Corbin, Kentucky.

7. In the 1930s, Sanders developed his secret recipe of 11 herbs and spices.

8. McDonald's was the most-valuable fast food brand in 2022, followed by Starbucks.

9. McDonald's was founded in the early 1940s by Dick and Mac McDonald as a barbecue drive-in.

10. The early McD's even offered carhop service.

11. French fries weren't introduced at Mickey D's until 1949.

12. Until then, potato chips had been offered instead.

13. McDonald's has its own university (of sorts).

14. Hamburger University opened in 1961.

15. There graduates receive "Bachelor of Hamburgerology" degrees.

16. Ronald McDonald first appeared in 1966 when McDonald's aired its first television commercial.

17. The Hamburglar, Grimace, and Mayor McCheese joined him five years later.

18. Insta-Burger King was founded in Jacksonville, Florida, in 1953.

19. James McLamore and David Edgerton purchased the company in 1954 and renamed it "Burger King."

20. The restaurant was based on an assembly line production system inspired by a visit to the McDonald brothers' hamburger stand.

21. Today, Burger King has more than 17,000 locations worldwide.

22. Australia is the only country in which Burger King does not operate under its own name.

23. Burger King is known as Hungry Jack's in Australia.

24. Burger King says there are 221,184 possible ways you could order its Whopper hamburger.

25. Taco Bell comes from a family name: Glen Bell started working on the chain in the late 1940s in San Bernardino—the same place where McDonald's was born.

26. His first venture was a hot dog stand called Bell's Drive-In.

27. By the early '50s, Bell started adding Mexican items onto the menu, eventually opening a secondary restaurant called Taco Tia.

28. The first actual Taco Bell didn't open until 1962.

29. In 1958, brothers Dan and Frank Carney of Wichita, Kansas, founded Pizza Hut.

30. With more than 18,000 restaurants, Pizza Hut is today the world's largest pizza chain in terms of locations.

31. Brothers Tom and James Monaghan purchased a pizza store in Ypsilanti, Michigan, called DomiNick's for $500 in 1960.

32. A year later, Tom became the restaurant's sole owner when James traded his share of the business for a Volkswagen Beetle.

33. Tom renamed the store Domino's Pizza and it soon became one of the world's leading pizza chains.

34. Today, Domino's is the top pizza chain in the world in terms of revenue.

35. Arby's entered the restaurant world in 1964.

36. The first location opened in Boardman, Ohio, featuring the roast beef sandwiches that are still the chain's signature item today.

37. The name Arby's actually represents the initials "R" and "B."

38. The letters stand for "Raffel Brothers," in homage to founders Leroy and Forrest Raffel, although the company says many suspect it also stood for "roast beef."

39. Subway was founded in 1965 by 17-year-old college freshman Fred DeLuca and family friend Dr. Peter Buck.

40. If all the sandwiches made by Subway in a year were placed end to end, they would wrap around the world an estimated six times.

41. Subway's Italian B.M.T., is named after the Brooklyn Manhattan Transit.

42. Wendy's joined the fast food mix in 1969 when founder Dave Thomas opened his first restaurant in Columbus, Ohio.

43. Wendy's was named for Dave Thomas's daughter, Melinda Lou "Wendy" Thomas.

44. McDonald's Filet-O-Fish was originally developed specifically for Catholic customers abstaining from eating meat on Fridays.

45. Steve Ellis founded Chipotle in 1993 after graduating from culinary school in order to fund his dream of opening a fine-dining restaurant.

46. According to the book *Eat This, Not That*, Chick-fil-A is the healthiest overall fast food chain.

47. Chick-fil-A was the first fast-food chain in the United States to offer a menu completely free of trans fat.

48. Shaquille O'Neal owns some 155 Five Guys restaurants in North America.

20 Pasta Translations

1. Tortellini = twists
2. Vermicelli = worms
3. Spaghetti = strings or twine
4. Farfalle = butterflies
5. Fettuccine = little ribbons
6. Fusilli = springs or rifles
7. Linguine = little tongues
8. Manicotti = muffs or sleeves
9. Mostaccioli = mustaches
10. Penne = pens or quills
11. Rotelle = little wheels
12. Ziti = bridegrooms
13. Campanelle = little bells
14. Ditali = thimbles
15. Conchiglie = shells
16. Lasagna = flat sheets
17. Barbina = little beards
18. Ditalini = small fingers
19. Reginette = little queens
20. Lumaconi = snails

Chapter 6

SPORTS AND GAMES

25 Early Sports & Games Facts

1. The ancient Egyptians loved playing a checkerboard game called senets.

2. Players moved pieces around a board determined by tossing numbered throwing sticks.

3. Chinese acrobatics were performed at least as far back as the Han Dynasty, around 200 B.C.

4. Central American Mayans developed their own team sports similar to lacrosse, football, and soccer.

5. Ancient Egypt had its own version of the Olympics, featuring gymnastics, javelin, running, swimming, and other events.

6. Native American tribes had many versions of, and names for, the modern sport of lacrosse.

7. The Cherokee called it "Little Brother of War," because it was good training for combat.

8. The Roman game of quoits, where a ring is tossed at a stake in the ground, is the forerunner of modern horseshoes.

9. An ancient festival devoted to cheese rolling is still held each May in Gloucestershire, England.

10. The festival dates back hundreds of years and involves pushing and shoving a large wheel of ripe Gloucestershire cheese downhill in a race to the bottom.

11. Skipping and jumping are natural movements of the body, and the inclusion of a rope in these activities dates back at least as far as A.D. 1600, when Egyptian children jumped over vines.

12. Early Dutch settlers brought the game of jump rope to North America.

13. It flourished and evolved from a simple motion into the often-elaborate form prevalent today: double Dutch.

14. With two people turning two ropes simultaneously, a third, and then fourth, person jumps in, often reciting rhymes.

15. Ice skates were used as far back as the Bronze Age, although early skates were likely used for winter travel, not for fun.

16. Something like today's Hula-Hoop existed some 3,000 years ago, when children in ancient Egypt made hoops out of grapevines and twirled them around their waists.

17. The first historical mention of the yo-yo dates to Greece in 500 B.C., but it was a man named Pedro Flores who brought the yo-yo to the United States from the Philippines in 1928.

18. The original name of the game volleyball was "mintonette."

19. It was created in 1895 when a YMCA gym teacher borrowed from basketball, tennis, and handball to create a new game.

20. Playing cards have been a tradition in China for millennia.

21. The markings on cards today date back to 14th-century France.

22. The four suits represent the major classes of society at that time.

23. Hearts translate to nobility and the church; the spear-tip shape of spades stands in for the military; clubs are

clover, meant to represent rural peasantry; and diamonds are similar to the tiles then associated with retail shops and signified the middle class.

24. The classic territorial-capture board game Go originated in China some 4,000 years ago.

25. Legend has it that an emperor invented it to sharpen his son's thinking.

17 Ancient Olympic Games Facts

1. The Games took place at Olympia, a religious sanctuary.

2. The nearest town was a small one called Elis that was about 40 miles away.

3. Reportedly the Games began in 776 B.C.

4. At the time, Greece consisted of city-states, and Rome was not yet an empire.

5. Legend has that the King of Elis held the Games for the first time after an Oracle told him it would appease Zeus and end a plague.

6. Elis continued to host the Games as a multi-day event every few years.

7. Over time, training facilities and arenas were built to accommodate the events, which grew larger over time.

8. Footraces, equestrian, and combat events all took place.

9. Any male athlete could try out.

10. The Games were winner-take-all—no silver or bronze medals.

11. Married women were not permitted to compete nor attend.

12. One exception to this rule held that female racing-chariot owners were allowed.

13. A priestess of Demeter was also permitted to attend.

14. *Pankratists*—freestyle fighters—and other athletes including charioteers and boxers, literally risked their health and lives.

15. Injuries and deaths were not uncommon.

16. The audience flocked to the Games despite a lack of sanitation and the risk of sunstroke.

17. The Games were banned by Roman Emperor Theodosius I in A.D. 393 to promote Christianity.

32 Facts About Golf

1. King James II of Scotland banned golf in 1457 because golfers spent too much time on the game instead of improving their archery skills.

2. The first golf course in England was the Royal Blackheath Golf Club, founded in 1608.

3. The first golf course constructed in the United States was Oakhurst Links Golf Club in White Sulfur Springs, West Virginia, in 1884.

4. It was restored in 1994, but then closed to repair flood damage from 2016.

5. The oldest continuously existing golf club in the United States is St. Andrews Golf Club in New York.

6. It was formed by the Apple Tree Gang in 1888.

7. The first golf balls were made of wood.

8. Next were leather balls filled with goose feathers, followed by rubber balls, gutta-percha balls (made of a leathery substance from tropical trees), and then modern balls.

9. The first sudden-death playoff in a major championship was in 1979, when Fuzzy Zoeller beat Tom Watson and Ed Sneed in the Masters.

10. The longest sudden-death playoff in PGA Tour history was an 11-hole playoff between Cary Middlecoff and Lloyd Mangrum in the 1949 Motor City Open.

11. They were declared co-winners.

12. Beth Daniel is the oldest winner of the LPGA Tour.

13. She was 46 years, 8 months, and 29 days old when she won the 2003 BMO Financial Group Canadian Women's Open.

14. Most golf courses in Japan have two putting greens on every hole—one for summer and another for winter.

15. Tiger Woods has won eleven PGA Player of the Year Awards as of the 2022 season, the most won by any PGA Tour player.

16. Tom Watson is second, with six awards.

17. The youngest player to win a major championship was Tom Morris Jr. (known as "Young" Tom), who was 17 years old when he won the 1868 British Open.

18. Sam Snead and Tiger Woods top the list of PGA Tour winners with 82 victories each as of 2022.

19. They are followed by Jack Nicklaus with 73 wins, Ben Hogan with 64, and Arnold Palmer with 62.

20. To make the LPGA Hall of Fame, players must be active on the tour for ten years, have won an LPGA major championship, and secured a significant number of trophies, among other qualifications.

21. The shortest hole played in a major championship is the 106-yard, par-3 seventh hole at Pebble Beach Golf Links in California.

22. The longest hole in the United States is the 841-yard, par-6 12th hole at Meadows Farms Golf Club Course in Locust Grove, Virginia.

23. In 1899, Dr. George Grant, a New Jersey dentist, invented and patented the first wooden tee.

24. Before tees were invented, golfers elevated balls on a tiny wet-sand mound.

25. The oldest player to win a major championship was 48-year-old Julius Boros at the 1968 PGA Championship.

26. The youngest golfer to shoot a hole-in-one was five-year-old Coby Orr.

27. It happened in Littleton, Colorado, in 1975.

28. The chances of making two holes-in-one in a single round of golf are 1 in 67 million.

29. There are more than 11,000 golf courses in North America.

30. Playing on a downhill hole and with the help of a tail-wind, Bob Mitera sank a 444-yard hole-in-one on the Miracle Hill Country Club course in Omaha, Nebraska.

31. When you score two shots under par, it's called an eagle.

32. When you shoot three under par on a single hole, it's called an albatross.

14 Great Sports Nicknames

1. One of hockey's tougher pugilists, 6'5" Stu "The Grim Reaper" Grimson earned more than 2,100 penalty minutes during his 729-game NHL career.

2. Basketball Player "Pistol Pete" Maravich used his eerie peripheral vision to pull off hotdog passes and circus shots like one of the Harlem Globetrotters.

3. During Wayne Gretzky's heyday with the great Edmonton teams, Dave "Cementhead" Semenko had one job: Keep Wayne safe.

4. Football player Dick "Night Train" Lane reportedly got the odd nickname from associating with fellow Hall-of-Famer Tom Fears, who constantly played the record *Night Train* on his phonograph.

5. Basketball player Damon Stoudamire stood only 5'10". Between this and a tattoo of "Mighty Mouse" on his arm, the nickname "Mighty Mouse" was inevitable.

6. Football player for the Chicago Bears and the Philadelphia Eagles, William "The Refrigerator" Perry got attention for Chicago coach Mike Ditka's willingness to use the 326-pound "Fridge" at fullback on goal-line plays.

7. The last pitcher legally allowed to throw the spitball under the grandfather clause when baseball outlawed ball-doctoring, Burleigh "Ol' Stubblebeard" Grimes always showed up to the diamond with a faceful of scruffy whiskers.

8. Baseball player Mike "The Human Rain Delay" Hargrove got his nickname by fooling around in the batter's box: He would adjust his helmet, adjust his batting glove, pull on his sleeves, and wipe his hands on his pants before every pitch.

9. Basketball player Darrell "Dr. Dunkenstein" Griffith of the Utah Jazz was only 6'4" but could jump as though grafted to a pogo stick.

10. Pepper "The Wild Horse of the Osage" Martin was a player for the St. Louis Cardinals in the 1930s and '40s. His rather wild, free-spirited base-running got him the nickname, though it also probably referred to his love of practical jokes. "Pepper" was a nickname too: he was born Johnny Leonard Roosevelt Martin.

11. Basketball star Charles Barkley's nicknames included "The Round Mound of Rebound."

12. Norwegian biathlete Ole Einar Bjorndalen is known as the "King of Biathlon," which makes sense. He's also

known as "The Cannibal," leading to a number of head-lines about him being "hungry for success" and (after a defeat) "toothless."

13. With the last name "Crawley," it was probably inevi-table that British Cricketeer John Crawley would be dubbed "Creepy."

14. Speaking of inevitable: Usain Bolt, of course, is known as Lightning Bolt.

17 Facts About Curling

1. Curling began in Scotland in the 1500s, played with smooth river-worn stones.

2. The game is also called "The Roaring Game" because of the noise the rocks make on the ice.

3. The Royal Montreal Curling Club began in 1807.

4. In 1927, Canada held its first national curling championship.

5. Curling was played as a demonstration sport in the Olympic Games in 1932, 1988, and 1992, and officially added in 1998.

6. A curling tournament in called a bonspiel.

7. A "cashspiel" is slang for a tournament with large prizes.

8. The standard curling rink measures 146 feet by 15 feet.

9. At each end are 12-foot-wide concentric rings called houses, the center of which is the button.

10. There are four curlers on a team.

11. Each shoves two rocks in an effort to get as close to the button as possible.

12. The players with the brooms aren't trying to keep the ice clear of crud.

13. The team skip (captain) determines strategy and advises the players using the brooms in the fine art of sweeping to guide the stone with surprising precision.

14. One particular arrangement of rocks in the house is called a Christmas Tree for its shape.

15. The hog line is a blue line in roughly the same place as a hockey blue line.

16. One must let go of the rock before crossing the near hog line—and the rock must cross the far hog line—or it's hogged (removed from play).

17. Good sportsmanship is prized, and players are even expected to call their own fouls.

13 Sports Salaries Facts

1. In the National Basketball Association's first season, 1946–47, the top-paid player was Detroit's Tom King, who made $16,500.

2. King also acted as the team's publicity manager and business director.

3. Players who ran the floor in the first professional basketball league, which was formed in 1898, were paid $2.50 for home games and $1.25 for road matches.

4. As a young man, legendary coach Vince Lombardi coached football, basketball, and baseball at Cecilia High School in Englewood, New Jersey, while also teaching Latin, algebra, physics, and chemistry.

5. He did all of this for $1,700 a year.

6. Hank Greenberg was the first baseball player to earn $100,000 a year in 1947.

7. In 1930, Babe Ruth signed a contract that would pay him $80,000 a year—a shocking sum that earned him more than then-President Herbert Hoover.

8. Ruth famously quoted, "Why not? I had a better year than he did."

9. In 1975, the late Brazilian soccer legend Pelé joined the North American Soccer League team the New York Cosmos for a record salary of $1.4 million per year.

10. In 2012, then-undefeated boxer Floyd Mayweather was ranked by *Forbes* as the world's highest-paid athlete.

11. He competed in two pay-per-view fights, earning $85 million in the process.

12. Earl Anthony was the first man to accumulate $1 million in career earnings as a professional bowler.

13. In 2017, *Forbes* produced a list of the 100 highest-paid athletes. The only woman on the list was Serena Williams, who made about $27 million that year.

24 Early Football Facts

1. The University of Toronto played the first documented game of something rugby-like that was potentially the first football game in the world in 1861.

2. In 1869, Princeton traveled to Rutgers for a rousing game of "soccer football."

3. The field was 120 yards long by 75 yards wide, about 25 percent longer and wider than the modern field.

4. It played more like soccer than modern football, and with 25 players on a side, the field was a crowded place. Rutgers prevailed 6–4.

5. In 1892, desperate to beat the Pittsburgh Athletic Club team, Allegheny Athletic Association leaders created the professional football player by hiring William "Pudge" Heffelfinger to play for their team.

6. Heffelfinger played a pivotal role in AAA's 4–0 victory.

7. The first Rose Bowl was played in 1902.

8. Michigan, having scored its regular schedule 501–0, drubbed Stanford 49–0.

9. In 1902, Charles Follis joined the Shelby Blues, a team in Ohio, becoming the first Black professional football player.

10. Follis later went to play baseball for the Cuban Giants, the first Black professional baseball club.

11. One of Follis's football teammates, Branch Rickey, also went on to a career in baseball. Rickey later signed Jackie Robinson to the Dodgers.

12. In 1905, disgusted at the mortality rate among college football players, Teddy Roosevelt told the Ivy League schools: "Fix this blood sport, or I'll ban it."

13. A number of changes followed, including a ban on mass plays.

14. The mortality rate went from 19 deaths in 1905 to 11 in the following year.

15. In 1911, in the first championship of the National Football League (NFL), the Pittsburgh Steelers defeated the Philadelphia Athletics by a score of 11 to 0.

16. In 1912, the Rules Committee determined that a touchdown is worth six points.

17. In 1921, fans heard the first commercially sponsored radio broadcast of a game, with University of Pittsburgh beating West Virginia 21–13.

18. In 1922, the American Professional Football Association became the National Football League (NFL).

19. The NFL didn't begin recording statistics until 1932.

20. The modern football took its current shape in the mid-1930s, after a couple of decades of gradual evolution from the egglike rugby ball.

21. In 1939, the Brooklyn Dodgers–Philadelphia Eagles game was the first to be beamed into the few New York homes that could afford TV sets.

22. The first coast-to-coast TV broadcast of an NFL game was in 1951 as the Los Angeles Rams faced the Cleveland Browns in the league championship game.

23. The American Football League (AFL), the NFL's rival, began play in 1960.

24. The two merged in 1970.

12 College Football Facts

1. Legendary Notre Dame football coach Knute Rockne won more than 88 percent of his games, with a 105–12–5 career.

2. Ohio State running back Archie Griffin was the first player to win the Heisman Trophy twice, in 1974 and 1975.

3. The first freshman to win the Heisman was Texas A&M quarterback Johnny Manziel in 2013.

4. Lafayette and Lehigh in Eastern Pennsylvania have been famous football rivals since 1884.

5. Professional basketball player Charlie Ward played football in college for Florida State.

6. The University of Nebraska–Lincoln sold out a game against Missouri in 1962—and just kept selling out.

7. The Cornhuskers didn't have an unsold seat for 375 games.

8. The Michigan Wolves have won more Division I games than any other team.

9. Toledo won the first overtime game in major college football history in 1995, defeating Nevada in the Las Vegas Bowl.

10. The Colorado Buffaloes have been running onto the field behind their mascot "Ralphie," a real live buffalo, since 1967.

11. There have been six Ralphies, and despite the name, they have all been female.

12. "Dotting the i" in the "Script Ohio" is one of the most memorable traditions among marching bands nation-wide and is considered to be a high honor.

29 Tennis Facts

1. The first reliable accounts of tennis come from tales of 11th-century French monks who played a game called *jeu de paume* ("palm game," that is, handball) off the walls or over a stretched rope.

2. The main item separating tennis from handball—a racket—evolved within these French monasteries.

3. Rackets had been used before that in ancient Greece, in a game called *sphairistike* and then in *tchigan*, played in Persia.

4. Henry VII popularized the game in England.

5. In 1529 Henry VII built a tennis court on the grounds of his Hampton Court palace.

6. It was rebuilt in 1625 and is now the oldest tennis court in existence.

7. Thirty cases of golf clubs and tennis rackets for A. G. Spalding's sporting goods company were being shipped by transatlantic mail on the *Titanic* when it sank.

8. That wasn't the only *Titanic* tennis tie: Known for his victories as a singles and doubles tennis player at the U.S. Open and Wimbledon during the 1910s and '20s, R. Norris Williams also gained fame as a survivor of the *Titanic* disaster.

9. Six competitors from four nations participated in the women's singles event at the 1900 Olympics, the second in the modern era.

10. The first female Olympic champion was Charlotte Cooper of Great Britain, who won the tennis singles and the mixed doubles.

11. For his habit of indulging temper tantrums, John McEnroe was once dubbed by *The New York Times* "the worst advertisement for our system of values since Al Capone."

12. Wimbledon drama! In 1995, tennis player Jeff Tarango disagreed vehemently with the chair umpire.

13. His wife Benedictine came onto court and slapped the official. The two then stormed off the court.

14. Drama in tennis is nothing new. While playing a game of tennis in 1606, the artist Caravaggio and his young opponent got into an argument.

15. Caravaggio stabbed and killed his opponent and immediately had to flee Rome.

16. Catgut (the name of the product which tennis racket strings are made of) is made from the intestines of mostly livestock animals, but not cats.

17. In 2002, *Sports Illustrated* ran a profile of a 17-year-old from Uzbekistan, Simonya Popova.

18. She was everything a rising tennis star would want to be, except for the fact that she didn't exist: The article was a spoof.

19. Between 1927 and 1932, four French tennis players who were dubbed "the Four Musketeers" won six consecutive Davis Cups, as well as numerous Wimbledon and French championships.

20. In 1973, Billie Jean King defeated Bobby Riggs in three straight sets in a tennis match billed as the "Battle of the Sexes."

21. Sir Elton John wrote a song for Billie Jean King, titled "Philadelphia Freedom," that was released in 1975.

22. According to most tennis historians, modern tennis dates back to the early 1870s, when the delightfully named Major Walter Clopton Wingfield devised a lawn game for the entertainment of party guests on his English country estate.

23. Wingfield (whose bust graces the Wimbledon Tennis Museum) based his game on an older form of tennis that long had been popular in France and England, called "real tennis."

24. Wingfield opted to borrow the counting system from earlier versions of tennis—in French, scoring mimicked the quarter-hours of the clock: 15–30–45.

25. For some unknown reason, 45 became 40.

26. A number of historians argue that Wingfield borrowed the terms for his new game from the older French version.

27. Hence, *l'oeuf* (meaning "egg") turned into "love" and *deux le jeu* ("to two the game") became "deuce."

28. Actor Brad Pitt played tennis in high school at Kickapoo High in Springfield, Missouri.

29. In 1957, tennis great Althea Gibson won Wimbledon, the first African American player to do so.

25 Facts About Jim Thorpe

1. Generally considered one of the finest athletes of all time, James Thorpe played baseball and football at a professional level.

2. He also competed in the Olympics as a runner in the decathlon and pentathlon events.

3. Born in 1887 or 1888 in what is now Oklahoma, Thorpe was Native American, a member of the Sac and Fox Nation.

4. His Native American name translates to "Bright Path."

5. As a teenager, he was coached in football by legendary Glenn "Pop" Warner.

6. Thorpe also competed in track and field events, baseball, and lacrosse as a teenager.

7. He led his team to victory in 1912 in the national collegiate championship.

8. He competed in track and field events in the 1912 Summer Olympics in Stockholm, Sweden.

9. Thorpe won gold in the pentathlon and decathlon.

10. He won eight of the 15 individual events that comprised the two competitions.

11. Thorpe was honored on his returned home with a ticker-tape parade on Broadway.

12. After the Olympics, Thorpe continued to play various sports.

13. In 1913, a story broke that Thorpe had played semi-professional baseball in 1909 and 1910.

14. Since he had not been an amateur, he was stripped of his Olympic medals.

15. Because the reports did not come out until a year later, Thorpe's disqualification was actually against the rule-book for the 1912 Olympics, which stated that protests had to be made within 30 days.

16. Because of this, and because it wasn't uncommon for college players to earn money playing during the

summer (often using an alias), Thorpe's disqualification was perceived as unfair.

17. Thorpe's medals were reinstated in a ceremony in 1983 after a long campaign by people who felt he had been treated unfairly and that racism might have been part of the disqualification.

18. His children received his reinstated medals on his behalf.

19. Thorpe played professional baseball in both the major and minor leagues.

20. He also played for several different NFL teams.

21. He was part of a traveling basketball teams that barnstormed the country in 1927 and 1928.

22. Thorpe appeared in several films, playing a football coach in *Always Kickin'* and an extra in Westerns.

23. Thorpe was the subject of a 1951 movie starring Burt Lancaster: *Jim Thorpe – All American*.

24. Thorpe struggled with financial issues and alcoholism later in life.

25. He died in 1953 at the age of 65.

21 Baseball Firsts

1. In 1882, Paul Hines became the first ballplayer to wear sunglasses on the field. They weren't corrective; he just didn't like having the sun in his eyes.

2. On July 4, 1939, Lou Gehrig was the first major-league player to have his number (4) retired.

3. On May 23, 1901, Nap Lajoie was the first player in baseball history to be intentionally walked with the bases loaded.

4. During a spring training game on March 7, 1941, Pee Wee Reese and Joe Medwick of the Brooklyn Dodgers

were the first major-league players to wear plastic batting helmets.

5. The first night game in World Series history was Game 4 of the 1971 series, when Pittsburgh hosted Baltimore.

6. The cork center was added to the official baseball in 1910. Before that, the core of a baseball was made of rubber.

7. In 1957, Warren Spahn became the first left-handed pitcher to win a Cy Young Award.

8. In 1953, respected and innovative National League umpire Bill Klem was the first ump elected to baseball's Hall of Fame.

9. On May 17, 1939, the first televised baseball game took place. It was a college game between Princeton and Columbia broadcast on W2XBS. Princeton won 2–1.

10. The first All-Star Game broadcast on television took place on July 11, 1950.

11. On April 18, 1956, Ed Rommel was the first umpire to wear glasses in a regular-season game.

12. Hall of Fame player Frank Robinson was the first Black manager in the majors, but the Blue Jays' Cito Gaston was the first Black manager whose team won the World Series.

13. The first "Babe" in baseball was Babe Adams, who pitched from 1906 to 1926.

14. The first major-leaguer to hit a home run with the lights on was Babe Herman of the Reds in July 1935.

15. Babe Ruth hit his first career home run at the Polo Grounds against the New York Yankees on May 6, 1915.

16. The American League's first Most Valuable Player (MVP) Award was given to St. Louis Browns first baseman George Sisler in 1922. The second was given to Babe Ruth.

17. In 1928, the Hollywood Stars of the Pacific League became the first team to travel by air.

18. In 1981, Rollie Fingers of the Milwaukee Brewers became the first relief pitcher to win the American League MVP Award.

19. On April 23, 1952, Hoyt Wilhelm won his first game and hit his first home run, which became the only home run he ever hit in 1,070 games.

20. In 1960, Bill Veeck of the White Sox became the first owner to put players' names on the back of their uniforms.

21. In 1992, the Toronto Blue Jays won the World Series. In doing so, they became the first Major League Baseball team from a country other than the United States to win it.

20 World Series Facts

1. The modern World Series was first played in 1903.

2. Before 1903, a number of 19th-century competitions were held between various leagues. These series were sometimes called the "World's Series."

3. In the 1903 series, the Boston Americans (the predecessor team to the Red Sox) of the American League beat the Pittsburgh Pirates of the National League.

4. The World Series was not held in 1904 due to a boycott.

5. Other than that, the competition has been held each year except 1994, when there was a player's strike.

6. An earthquake interrupted Game 3 of the 1989 World Series between the San Francisco Giants and the Oakland Athletics.

7. Candlestick Park had to be evacuated and the game postponed.

8. The first trio of brothers ever to win World Series titles were the Molina brothers.

9. Bengie, the oldest, and Jose won championships as Los Angeles Angels teammates in 2002, and younger brother Yadier joined the club with the St. Louis Cardinals in 2006 and '11—all as catchers.

10. Sparky Anderson was the first manager to win World Series titles in both leagues.

11. He captured his first two championships with the Cincinnati Reds in the National League.

12. Anderson then took the Detroit Tigers of the American League to the top in 1984.

13. With 18 World Series home runs, Mickey Mantle holds the record for most career home runs in World Series play.

14. He topped Babe Ruth's previous record by three.

15. Reggie Jackson earned the nickname "Mr. October" in part by hitting home runs on four straight swings of the bat in the 1977 World Series.

16. Yogi Berra played in the World Series a record 14 times.

17. In 2000, Derek Jeter became the first in Major League history named MVP of both the All-Star Game and World Series in the same year.

18. "Shoeless" Joe Jackson went 12-for-32 (.375) during the 1919 World Series—the one for which he and seven Chicago White Sox teammates were banned for life for their roles in "fixing" games.

19. The New York Yankees have made it to the World Series 40 times, winning 27 times.

20. The Seattle Mariners have never played in a World Series.

43 Basketball Facts

1. In 1936, basketball was added as an Olympic sport.

2. In 2004, the University of Connecticut won both the men's and the women's NCAA basketball championships.

3. It is the only school to ever do the double dip.

4. The silhouette of a dribbling basketball player on the National Basketball Association's logo is an image of former Los Angeles Lakers great Jerry West.

5. There are numerous rules on how to properly dribble a basketball, but bouncing the ball with such force that it bounds over the head of the ball handler is not illegal.

6. When Dr. James Naismith first drafted the rules for the game that eventually became known as basketball, the dribble wasn't an accepted method of moving the ball.

7. In the game's infancy, the ball was advanced from team-mate to teammate through passing.

8. Naismith played the game only twice because he felt he committed too many fouls.

9. He believed this was because his extensive experience in wrestling and football made physical contact come naturally to him.

10. Wilt Chamberlain holds a number of NBA records, including the unapproachable mark of 100 points in a single game.

11. Chamberlain also holds the record for 55 rebounds in a game.

12. During his 100-point game, Chamberlain converted 28 of his 32 free throw attempts.

13. He was a 51-percent career shooter from the free throw line, but hit an .875 rate in his signature game.

14. Although the term *dunk* was commonly used to describe the action of propelling a basketball through the hoop from above the rim, the phrase *slam dunk* was coined by the late Los Angeles Lakers announcer Francis "Chick" Hearn.

15. The colorful commentator also originated the terms *air ball* (ball that misses the entire backboard), *charity stripe* (foul line), and *finger roll* (rolling the ball off the fingertips).

16. George Mikan, a 6'10" pioneer in pro basketball who played for the Chicago American Gears and the Minneapolis Lakers through the 1940s and '50s, was probably the first man to use the dunk as an offensive weapon.

17. In 2002, Lisa Leslie became the first woman in the WNBA to dunk.

18. In 1993, Michael Jordan broke the hearts of fans by retiring from the Chicago Bulls.

19. He returned to basketball 17 months later.

20. The film *Hoosiers* was inspired by the Cinderella story of Milan High School—a school that, despite having only 73 male students, won the Indiana State Championship in 1954.

21. "Be tall, bask," is an anagram of basketball.

22. Before 1937, the basketball referee tossed a jump ball after every basket.

23. In 1895, the first American college basketball game was played between the Minnesota State School of Agriculture and Hamline College.

24. At the time, peach baskets were still used for hoops.

25. Minnesota State won the game, 9 to 3.

26. In 1946, the Basketball Association of America (BAA) was founded.

27. The BAA merged with the National Basketball League in 1949 to form the National Basketball Association (NBA).

28. In 2013, Rick Pitino became the first coach to lead two different schools to men's NCAA Division I championships.

29. UCLA had an incredible 88-game winning streak in 1974, broken by Notre Dame.

30. Lewis Alcindor, before he became known as Kareem Abdul-Jabbar and a professional player, was honored as Most Outstanding Player in the NCAA Final Four a record three times, in 1967, '68, and '69.

31. Michael Jordan launched a 17-foot winning shot in the 1982 NCAA Title Game that led North Carolina to victory over Georgetown.

32. Jordan was a freshman at the time.

33. Legendary coach Pat Summit won eight national titles during her tenure at Tennessee from 1974 to 2012.

34. Under Don Haskins in 1966, Texas Western College upset Kentucky to become the first NCAA men's basketball champion with an all-Black starting lineup.

35. Before they were famous, Magic Johnson and Larry Bird battled for the 1979 NCAA championship.

36. Michigan State's Magic Johnson prevailed over Indiana State's Larry Bird in what was then the highest-rated college basketball game in television history.

37. Michael Jordan led the Chicago Bulls to six NBA championships in the 1990s.

38. Jordan was MVP of all six of his NBA Finals appearances, a record.

39. LeBron James, as of 2022, has been named MVP four times.

40. Coach Phil Jackson has won a "three-peat" three different times!

41. Jackson led the Chicago Bulls to three straight titles twice in the 1990s, and then took the Los Angeles Lakers to three straight beginning with the 2000 championship.

42. On February 7, 2023, LeBron James scored his 38,388th point, breaking Kareem Abdul-Jabbar's all-time scoring record of 38,387 points.

43. The NBA career scoring record had stood for nearly 39 years.

16 Facts to Bowl You Over

1. Some historians trace bowling's roots back to 3200 B.C., while others place its origin in Europe in the third century A.D.

2. Legend has it that King Edward III banned bowling after his good-for-nothing soldiers kept skipping archery practice to roll.

3. Well into the 19th century, American towns were outlawing bowling, largely because of the gambling that went along with it.

4. Originally, it was a game of ninepin set up in a diamond formation.

5. The German immigrants (who popularized the game in the 1800s and saw it outlawed because of the gambling) added another pin, changed the formation to a triangle, and satisfied the courts that it was a different game. Modern bowling was born.

6. Bowling was originally outdoor fun, played in the sun.

7. In 1895, the American Bowling Congress (which is now known as the United States Bowling Congress) was formed, and local and regional bowling clubs began proliferating.

8. The first National Bowling Championship was held in Chicago, Illinois in 1901.

9. Frank Brill won the individual bowling championship with a score of 648.

10. In 1997, University of Nebraska sophomore Jeremy Sonnenfeld became the first bowler to "knock 900" by rolling three perfect 300 games in a row in an official tournament.

11. The world's largest bowling alley is found in Japan.

12. It takes pretty bad luck for someone to lose their bowling ball in their practice backswing during a professional event such as the Dayton Classic.

13. Doing it four times in one event will bring notice, as it did Fran Wolf in 1976. The last one got her an ovation!

14. In 1930, Wisconsinite Jennie Hoverson became the first woman to bowl a perfect game in the history of league bowling. Her recognition came much later, however.

15. African Americans were not allowed into sanctioned league bowling until 1951, four years after Jackie Robinson broke the color barrier in baseball.

16. A bowling pin needs to tilt at least 7.5 degrees to fall over.

34 Facts About Soccer

1. Evidence of games resembling soccer has been found in cultures that date to the third century B.C.

2. The Romans brought their version of the sport along when they colonized what is now England and Ireland.

3. On March 20, 1976, while playing for Britain's Aston Villa soccer team, footballer Chris Nicholl scored every goal in a 2–2 draw against Leicester City, including two "own goals," or goals for the opposing team.

4. Like golf, soccer (football) was outlawed in Scotland in 1457 because the sports were dangerous, time-wasting nuisances that detracted from more important pursuits—like archery.

5. During World War II, as Commonwealth forces prepared to make a stand at ancient Thermopylae in 1941, they held a scheduled soccer game and continued it even when strafed by Stukas.

6. Soccer is considered the most globally popular sport.

7. In 2003, Adidas erected a billboard in Tokyo's Shibuya district featuring two live human beings playing a game of "vertical soccer."

8. The soccer players were suspended from the billboard with bungee cords, as was the soccer ball.

9. In 2004, during an Olympic qualifying match between Peru and Argentina, frenzied Peruvian fans grew irate when referees disallowed a goal for the home team.

10. The resulting riot left 300 people dead and 500 injured.

11. The first radio broadcast of a soccer game was in 1927.

12. Teddy Wakelam provided play-by-play and commentary of a match between Arsenal F.C. and Sheffield United on BBC radio. The match was a 1–1 tie.

13. In 1958, an airplane carrying the Manchester United soccer club home from a European Cup playoff match crashed on takeoff at Munich airport.

14. Eight team members and 15 other passengers died in the crash.

15. In 1908, the soccer club Inter Milan was founded.

16. The club has played in the Italian top tier of professional soccer since its inception, winning the league title more than 17 times.

17. Founded in 1904, the Fédération internationale de football association (FIFA) is the international governing body of association soccer.

18. FIFA oversees international competition and its membership now includes 211 national associations.

19. The first FIFA World Cup was held in Uruguay in 1930.

20. Uruguay won the title match over Argentina, 4–2, in front of a crowd of more than 68,000 spectators.

21. Sheffield F.C. was founded in 1857.

22. The association football—or soccer—club in Sheffield, England, is the oldest still in operation.

23. In 1969, Brazilian soccer star Pelé scored his 1,000th goal.

24. The striker played for Santos Football Club from 1956–1974 and the New York Cosmos from 1975–1977, as well as the Brazilian national team.

25. During an unofficial Christmas truce on the Western front during World War I, the sides exchanged prisoners and food, sang Christmas carols together, and played games of football—or soccer—with one another.

26. The first FIFA Women's World Championship was held in China in 1991.

27. The U.S. women won the first World Cup.

28. With five men's titles as of 2023, Brazil has won more World Cups than any other country.

29. Contrary to belief, soccer balls are actually oval-shaped, not round.

30. The checkered pattern creates an illusion of a perfect sphere.

31. Referees were not used in official soccer matches until 1881.

32. The fastest red card in history was given to player Lee Todd in 2000 for using an expletive two seconds into the game.

33. A soccer field is called a "pitch," because regulation fields are pitched, or sloped, five degrees upwards from one end to the other.

34. The teams switch sides at half so each has to play slightly uphill.

16 Rocky Marciano Facts

1. Rocky Marciano was the heavyweight champion of the world from 1952 to 1956.

2. Marciano was the only heavyweight champion to retire without a defeat or a draw.

3. Rocky Marciano was born Rocco Marchegiano in 1923.

4. Marciano survived a near-fatal bout of pneumonia in childhood.

5. As a young adult lacking a modern gym, Marciano practiced boxing by hanging an old mailbag to a tree in his backyard.

6. He quit high school and worked as a ditchdigger and at a coal company.

7. Marciano served during World War II in the Army.

8. He won the 1946 amateur armed forces boxing tournament.

9. He wasn't undefeated as an amateur, finishing the year with an 11–3 record.

10. Marciano played baseball as a child and even tried out for the Chicago Cubs in 1947.

11. He was cut in three weeks, though, and turned professional in boxing.

12. In 1951, Marciano competed against Joe Louis in the last match of Louis's career. Marciano won.

13. Marciano defended his title of heavyweight champion six times.

14. Marciano's last bout took place in 1955 against Archie Moore.

15. The retired boxer died in 1969 in a private plane crash.

16. Marciano was honored with a commemorative postage stamp in 1999.

34 Popular Toys & Games Facts

1. Bingo started sometime around 1929 as "Beano."

2. A smart entrepreneur named Edwin Lowe spotted a crowd playing Beano at a fair and ran with the idea.

3. The first boxed Stratego set in the U.S. came out in 1961.

4. An ancient Chinese game called Jungle or Animal Chess very much resembles Stratego, and a Frenchwoman patented its modern incarnation as *L'attaque* in 1910.

5. The Milton Bradley Company released Twister in 1966, and it was the first game in history to use the human body as an actual playing piece.

6. A barber from Ohio invented the popular all-ages card game Uno in 1969.

7. Creator Merle Robbins sold Uno to a game company in 1972 for $50,000 plus royalties.

8. Mike Marshall created the Hacky Sack, a version of the footbag, in 1972 to help his friend John Stalberger rehabilitate an injured leg.

9. Enrico Rubik introduced his puzzle cube in 1974, and it became popular in the 1980s, confounding millions of

people worldwide with its 43 quintillion (that's 43 followed by 18 zeros) solutions.

10. In the early 1940s, a torsion spring fell off marine engineer Richard James's desk and tumbled end over end across the floor.

11. Since then, more than a quarter billion Slinkys have been sold worldwide.

12. The Super Soaker water gun was invented in 1988 by aerospace engineer Lonnie Johnson.

13. Silly Putty was developed in 1943 when James Wright, a General Electric researcher, was seeking a synthetic rubber substitute.

14. Silly Putty debuted as a toy in 1950.

15. Mr. Potato Head, with his interchangeable facial features, was patented in 1952.

16. Mr. Potato Head was the first toy to be advertised on television.

17. For the first eight years, parents had to supply children with a real potato until a plastic potato body was included in 1960.

18. Intending to create a wallpaper cleaner, Joseph and Noah McVicker invented Play-Doh in 1955.

19. Barbie came onto the toy scene in 1959, the creation of Ruth Handler and her husband Elliot.

20. Ruth and Elliot Handler, along with Harold Matson, founded the Mattel toy company.

21. Chatty Cathy, also released by the Mattel Corporation in 1959, was the era's second most popular doll.

22. Betsy Wetsy also made a splash with 1950s-era children.

23. Created by the Ideal Toy Company, Betsy's already-open mouth would accept a liquid-filled bottle.

24. Since 1963, when they were first introduced, more than 16 million Easy Bake Ovens have been sold.

25. A lightbulb provided the heat source for baking mini-cakes in America's first working toy oven.

26. Toy lovers have to salute manufacturer Hasbro for its G.I. Joe action figure, which first marched out in 1964.

27. The 11-inch-tall doll for boys had 21 moving parts.

28. Hot Wheels screeched into the toy world in 1968, screaming out of Mattel's concept garage with 16 miniature autos.

29. Weebles were released by Hasbro in 1971.

30. At the height of their popularity, the Weeble family had its own tree house and cottage.

31. Sweet-smelling Strawberry Shortcake was created in 1977 by Muriel Fahrion for American Greetings.

32. The company expanded the toy line in the 1980s to include Strawberry's friends and their pets.

33. Xavier Roberts was a teenager when he launched his Babyland General Hospital during the 1970s in Cleveland, Georgia, allowing children to adopt a "baby."

34. In 1983, the Coleco toy company began to mass-produce these dolls as Cabbage Patch Kids.

18 Facts About Pinball

1. Pinball was invented in the 1930s.

2. Pinball derived from the 19th-century game bagatelle.

3. Bagatelle involved a billiards cue and a playing field full of holes.

4. Some early pinball arcades "awarded" players for high scores.

5. In the mid-1930s, machines were introduced that provided direct monetary payouts.

6. This quickly earned pinball a reputation as a gambling device.

7. Starting in the 1940s, New York City Mayor Fiorello LaGuardia declared pinball parlors akin to casinos ("magnets for the wrong element"), ushering in an era of pinball prohibition.

8. Chicago, Los Angeles, and other cities followed, banning the game.

9. Despite or because of the ban, pinball became a favorite pastime among adolescents and teens in the 1950s.

10. Gottlieb's Humpty Dumpty, designed by Harry Mabs in 1947, was the first pinball game to feature flippers (three on each side) that allowed the player to use hand-eye coordination to influence gravity and chance.

11. In 1948, pinball designer Steven Kordek repositioned the flippers (just two) at the bottom of the playfield, and the adjustment became the industry standard.

12. New York's pinball embargo lasted until 1976.

13. City officials destroyed 11,000 machines before it was lifted.

14. The ban was lifted after council members voted 6–0 to legalize pinball in the Big Apple.

15. The turning point: Writer and pinball wizard Roger Sharpe called his shots during a demonstration in front of the New York City Council, proving that pinball was indeed a game of skill.

16. Though pinball popularity receded with the advent of the video game, it enjoyed the first of several revivals later in the 1970s, thanks to its association with such rock-and-roll luminaries as The Who, Elton John, and Kiss.

17. In 1991, Bally's introduced The Addams Family game to tie in with the release of the movie.

18. The Addams Family became one of the best-selling pinball games of all time, with 22,000 machines sold.

31 Auto Racing Facts

1. In 1961, Phil Hill was the first American to win the Formula One World Championship.

2. Mario Andretti was the second, in 1978.

3. Al Unser Sr. and Jr. were the first father-son duo to find victory lane at the Indianapolis 500.

4. Formula One driver Sebastian Vettel of Germany was just 23 years old when he became the youngest world champion in history in 2010.

5. The International Motorsports Hall of Fame was founded by NASCAR architect Bill France Sr. near Talladega Superspeedway in 1982.

6. The first running of the *24 Heures du Mans*, or 24 Hours of Le Mans, took place in 1923.

7. The most famous endurance race in the world will celebrate its 100th anniversary in 2023.

8. A.J. Foyt was the first driver to win the Daytona 500, Indy 500, 24 Hours of Daytona, and 24 Hours of Le Mans.

9. Foyt was also the first driver to win the Indy 500 in both a front-engine and rear-engine car.

10. In 2012, the Audi R18 E-Tron Quattro that won the 24 Hours of Le Mans was the first hybrid to win the event.

11. With 253 acres within its oval, the Indianapolis Motor Speedway, home of the Indianapolis 500, could house the Roman Colosseum, Churchill Downs, Yankee Stadium, Rose Bowl Stadium, and Vatican City.

12. Janet Guthrie, in 1977, was the first woman to qualify for and race in the Indy 500.

13. Mechanical problems led to an early exit from that race, but the following year she finished ninth.

14. The Indy 500 winner traditionally drinks milk in victory lane.

15. After three-time winner Louis Meyer drank buttermilk in victory lane in 1936, a dairy-industry executive made a pitch to keep the tradition going and it caught on.

16. Helio Castroneves was the first driver to take the checkered flag in his first two Indy 500 starts, in 2001 and 2002.

17. Honda provided the victorious engine for every Indy 500 win between 2004 and 2012.

18. Youngsters between the ages of 7 and 17 have been racing at the All-American Soap Box Derby, an annual summer festival based in Akron, Ohio, since 1935.

19. NASCAR's founding meeting was organized by Bill France Sr. on December 14, 1947, at the Streamline Hotel in Daytona Beach.

20. The circuit, called the "Strictly Stock" division, debuted two years later.

21. Richard Petty won ten consecutive NASCAR races in 1967, a NASCAR record that still holds as of 2023.

22. He won 27 of his 48 starts that year.

23. Jeff Gordon was the first NASCAR driver to host NBC's *Saturday Night Live*.

24. Danica Patrick became the first woman ever to win the pole for the Daytona 500 in 2013.

25. Driver Carl Edwards routinely performs a backflip to celebrate victories.

26. Jimmie Johnson won NASCAR's top series five years in a row from 2006 to 2010, the first (and so far only) NASCAR driver to win five consecutive Cup crowns.

27. Jeff Gordon was the fastest NASCAR driver in history to reach 50 career wins.

28. His win came in his 232rd race, the DieHard 500 at Talladega in 2000.

29. There was a tie in the NASCAR series points standings in 2011, requiring a tie-breaker to determine the Cup champion.

30. Tony Stewart and Carl Edwards were dead even in the final points. Stewart won his third Cup title based on total wins during the year.

31. The Netflix docuseries *Formula 1: Drive to Survive* has increased the popularity of international auto racing among an American audience.

21 Early Video Games Facts

1. At MIT in 1962, Steve Russell programmed the world's first video game on a bulky computer known as the DEC PDP-1.

2. Spacewar featured spaceships fighting amid an astronomically correct screen full of stars.

3. Nolan Bushnell founded Atari in 1972, taking the company's name from the Japanese word for the chess term "check."

4. Atari released the coin-operated Pong later that year.

5. Its simple, addictive action of bouncing a pixel ball between two paddles became an instant arcade hit.

6. In 1975, the TV-console version of Pong was released.

7. It was received with great enthusiasm by people who could play hours of the tennis-like game in the comfort of their homes.

8. After runaway success in the Soviet Union in 1985 (and in spite of the Cold War), Tetris jumped the Bering Strait and took over the U.S. market the next year.

9. Released in 1978, Midway's Space Invaders was a ubiquitous hit that generated a lot of money and also presented the "high score" concept.

10. A year later, Atari released Asteroids and outdid Space Invaders by enabling the high scorer to enter his or her initials for posterity.

11. The 1980 Midway classic Pac-Man was the world's most successful arcade game, selling some 99,000 units.

12. In 1980, Nintendo's first game Donkey Kong marked the debut of Mario, soon to become one of the most recognizable fictional characters in the world.

13. Originally dubbed Jumpman, Mario was named for Mario Segali, the onetime owner of Nintendo's warehouse in Seattle.

14. Designers originally wanted the title character of Q*bert (released 1982) to shoot slime from his nose.

15. But it was deemed too gross.

16. From 1988 to 1990, Nintendo sold roughly 50 million home-entertainment systems.

17. In 1996, Nintendo sold its billionth video game cartridge for home systems.

18. In 1981, 15-year-old Steve Juraszek set a world record on Williams Electronics' Defender.

19. His score of 15,963,100 got his picture in *Time* magazine—and it also got him suspended from school.

20. He played part of his 16-hour game when he should have been attending class.

21. Atari opened the first pizzeria/arcade establishment known as Chuck E. Cheese in San Jose in 1977.

23 Facts About Hockey

1. The NHL's "Original Six" franchises included the New York Rangers, Toronto Maple Leafs, Detroit Red Wings, Boston Bruins, Chicago Blackhawks, and Montreal Canadiens.

2. On November 1, 1959, goaltender Jacques Plante was the first goalie to wear a full protective mask.

3. He did so after taking a puck to the face that split his lip.

4. Clint Benedict had worn a half-mask for a brief time in 1930, but said it blocked his vision and scrapped it after a few games.

5. A goal, an assist, and a fight in the same game are said to comprise a "Gordie Howe hat trick."

6. Wayne Gretzy totaled 894 goals and 1,963 assists for 2,857 points during his illustrious career, the all-time NHL leader in those categories by far.

7. Coach Scotty Bowman holds the record for coaching wins, with 1,244 career victories.

8. Bowman was also the first coach to win Stanley Cups with three different teams: five with the Montreal Canadiens, and two each with the Pittsburgh Penguins and the Detroit Red Wings.

9. Goalie Martin Brodeur of the New Jersey Devils holds the record for most career victories for a goaltender, with more than 600.

10. Brodeur also holds the league's records for shutouts.

11. In 2012–13, the Chicago Blackhawks scored at least one point in 24 straight games to open the season, a feat unmatched in NHL history.

12. Dave "Tiger" Williams was a terror on the ice during the 1970s and 1980s, racking up 3,966 penalty minutes.

13. The NHL Winter Classic takes place outdoors. Brrrr!

14. Father and son Bobby and Brett Hull scored more than 1,350 goals between them.

15. Two years after retiring, Gordie Howe returned to the ice in his mid-40s, suiting up for the WHA's Houston Aeros with his son. He won the WHA scoring title, too.

16. In the 2006–07 season, 20-year-old Sidney Crosby became the youngest player to win a scoring title.

17. Crosby was also the youngest player to reach 200 career points.

18. The Montreal Canadiens hold the record for the franchise that has won the Stanley Cup the most times, with 24 wins as of 2022.

19. Lord Stanley of Preston, then the Governor General of Canada, paid $50 for the Stanley Cup in 1893.

20. In the eight seasons goaltender Ken Dryden played for the Montreal Canadiens, he won six Stanley Cups.

21. The Los Angeles Kings won their first ten road games of the 2012 Stanley Cup playoffs, a playoff record.

22. The first woman to have her name inscribed on the Stanley Cup was Detroit Red Wings president Marguerite Norris in 1955.

23. Chris Chelios played for the Stanley Cup in 24 different seasons with three different teams.

Chapter 7

SPACE

39 Solar System Facts

1. Humans have known about Mercury, Venus, Mars, Jupiter, and Saturn since prehistoric times because these planets are visible to the naked eye.

2. The word and original definition of "planet" are derived from the Greek *asteres planetai*, which means "wandering stars."

3. Planets are known as wanderers because they appear to move against the relatively fixed background of the stars, which are much more distant.

4. Stars and planets can be differentiated by two characteristics: what they're made of and whether they produce their own light.

5. Unlike stars, planets are built around solid cores.

6. They're cooler in temperature, and some are even home to water and ice.

7. The *Voyager 1* probe left Earth in 1977.

8. It (probably) reached interstellar space in 2013.

9. By NASA estimates, the heliosphere ends about 100 AU away from the sun, where an AU (astronomical unit) is the distance from Earth to the sun.

10. It's around 11 billion miles.

11. Neptune is about 30 AU away from the sun.

12. The Kuiper Belt, a ring containing icy objects, extends about 50 AU away.

13. The Kuiper Belt has been seen by the *New Horizons* probe.

14. Scientists believe our solar system formed 4.6 billion years ago.

15. Most of the mass of our solar system—more than 99 percent of it—is in the sun.

16. In fact, the mass of our solar system is "1.0014" solar masses, where one solar mass is the mass of the sun.

17. The nearest known star is Proxima Centauri, 4.25 light-years away.

18. There are objects in the solar system called centaurs.

19. Similar to asteroids and comets, centaurs have unstable orbits that cross the paths of one or more giant planets.

20. The Oort Cloud at the very edge of the solar system is named after Dutch astronomer Jan Oort.

21. Long-range period comets with orbits of longer than 200 years come from the Oort Cloud, whereas short-period comets come from the closer Kuiper Belt.

22. The term solar system appeared around 1700 in English.

23. Directives from the universe? "Try molasses" and "Slam oysters" are both anagrams of "solar system."

24. An asteroid belt lies between the rocky planets and the gaseous ones, located between Mars and Jupiter.

25. The four largest asteroids in this Main Asteroid Belt are Ceres (a dwarf planet), Vesta, Pallas, and Hygiea.

26. Scientists estimate that there are at least 1.1 million asteroids that measure a mile in diameter in the Main Asteroid Belt, and even more asteroids that are smaller than that.

27. The plane of Earth's orbit, and most of the other planets, is called the ecliptic.

28. Comets, however, orbit the sun at different angles.

29. All the planets in the solar system orbit the sun in the same direction.

30. Most other solar objects do as well, but some comets do not.

31. Halley's comet is one example of a comet with a retrograde orbit.

32. Halley's comet, a short-period comet, is visible from Earth every 75 years.

33. It's the only comet that we can see without a telescope that appears twice in a human life span.

34. Halley's comet will next appear in 2061.

35. During the formation of the solar system, there may have been hundreds of protoplanets.

36. Some joined together to become larger masses. Some were destroyed.

37. The objects we're familiar with—planets, asteroids, and so forth—are all found in the interplanetary medium.

38. In most other solar systems, the star is orbited by medium-sized planets that are larger than our rocky planets and smaller than our gas giants.

39. According to scientists, gold exists on Mars, Mercury, and Venus.

15 Animals in Space Facts

The first occupied spacecraft did not carry a human being or even a monkey. Instead, scientists launched man's best friend.

1. The Soviet Union launched the first living creature into orbit on November 3, 1957.

2. That creature was a three-year-old stray dog named Laika.

3. Weighing just 13 pounds, Laika's calm disposition and slight stature made her a perfect fit for the cramped capsule of Sputnik II.

4. In the weeks leading up to the launch, Laika was confined to smaller and smaller boxes for 15 to 20 days at a time.

5. Laika was also fed a diet of a special nutritional gel to prepare her for the journey.

6. Sputnik II was a 250-pound satellite with a simple cabin, a crude life-support system, and instruments to measure Laika's vital signs.

7. When the Soviets announced that Laika would not survive her historic journey, the mission ignited a debate in the West regarding the treatment of animals.

8. Initial reports suggested that Laika survived a week in orbit.

9. It was revealed many years later that Laika only survived for roughly the first five hours.

10. The craft's life-support system failed, and Laika perished from excess heat.

11. Despite her tragic end, the heroic little dog paved the way for occupied spaceflight.

12. Between 1957 and 1966, the Soviets successfully sent 13 more dogs into space—and recovered most of them unharmed.

13. Dogs were initially favored for spaceflight over other animals because scientists believed they could best handle confinement in small spaces.

14. In 1959, the United States successfully launched two monkeys into space.

15. Named Able and Baker, the monkeys were the first of their species to survive spaceflight.

32 Facts About Space & the Cosmos

1. The cosmos contains approximately 50 billion galaxies.

2. If you could fly across our galaxy from one side to the other at light speed, it would take 100,000 years to make the trip.

3. If you attempted to count the stars in a galaxy at a rate of one every second, it would take about 3,000 years to complete the task.

4. Space smells like diesel fuel, barbecue, and hot metal.

5. Astronauts can't smell space while they're on a space-walk, of course, because they're in their suits. However, the scent clings to their gear.

6. The Milky Way is a spiral galaxy that formed approximately 14 billion years ago.

7. Our solar system is just a small part of it.

8. Most of the stars we can see are in our galaxy.

9. Our galaxy is about 100,000 light-years in diameter and 1,000 light-years thick.

10. The sun, along with Earth, is around 26,000 light-years from the center of our galaxy, halfway to the edge of the galaxy along the Orion spiral arm.

11. The Magellanic Clouds, Earth's nearest galaxies, are visible only in the Southern Hemisphere.

12. Both are irregular blobs that appear to the naked eye as fuzzy patches.

13. Visible to the naked eye and easily enjoyed with bin-oculars, Andromeda is the closest spiral galaxy (like our own Milky Way).

14. You can spot the Andromeda galaxy in a place with little light pollution in the Northern Hemisphere, ideally in November.

15. We watched the Crab Nebula blow itself apart (or rather, we watched the light reach us).

16. In A.D. 1054, Arab and Chinese astronomers noted a star visible during daytime—now it has puffed out a big gas cloud.

17. The Crab Nebula is found in Taurus (winter, mainly in the Northern Hemisphere), just above the tip of the lower "horn."

18. You can get a reasonable view of the cloud surrounding the star's wreckage with a four-inch telescope, but binoculars will reveal something, too.

19. Not visible to the naked eye but fascinating in pictures or through a big telescope, the Ring Nebula looks like a smoke ring surrounding a small star.

20. Just south of Vega in the constellation Lyra (summer, mainly in the Northern Hemisphere) is a little parallelogram of stars, and the Ring Nebula is in the middle of its bottom short side.

21. Take a good look at Vega (impossible to miss), because in 14,000 years it'll be our North Star again as Earth's axis cycles around.

22. The Coalsack Nebula is smack in the middle of the Milky Way and big enough to block out most of the "milk."

23. You have to get below the equator to see it, but its position just left of and below the Southern Cross makes it easy to spot all year round.

24. It's not at all hard to find the Orion Nebula (winter, Northern Hemisphere; summer, Southern Hemisphere) in the dagger of Orion's three-star belt, and you can

easily see the haze with binoculars—great through a small telescope.

25. A star cluster, the Pleiades are one of the highlights of binocular astronomy, easy to spot in Taurus (winter, Northern Hemisphere; summer, Southern Hemisphere).

26. Look for a little coffee-cup shape of blue-white stars that show up sapphire in light magnification.

27. Algol is an eclipsing binary star (a bright star eclipsed at intervals by a dimmer nearby star).

28. For folks on Earth, Algol seems to vary in brightness, and you can tell with the naked eye; visibility will go from easy to difficult.

29. The eclipses last several hours and occur every three days.

30. A planet bigger than Jupiter? TrES-4 is 1,400 light-years from Earth and nearly twice the size of Jupiter.

31. TrES-4 is a planetary oddball, though, because of its low density—which is about the same as balsa wood.

32. The asteroid Psyche 16 is made of metal, mostly iron and nickel—so much metal that some estimate it is worth more than the entire global economy.

27 Facts About the Sun

1. Every year the sun loses 360 million tons.

2. In about five billion years, the sun will become a red giant.

3. The diameter of the sun is about 109 times the diameter of Earth.

4. In miles, the sun's diameter is 864,000 miles.

5. About three-quarters of the sun's mass is made of hydrogen.

6. Approximately one-quarter is made of helium and small amounts of oxygen, carbon, neon, and iron.

7. At its core, the sun converts hydrogen into helium.

8. The energy from that process creates the sun's light and heat.

9. The sun is brighter than about 85 percent of stars in the Milky Way.

10. The light from the sun takes an average of 8 minutes and 19 seconds to reach Earth.

11. The sun has a surface temperature of about 5,800 kelvin.

12. That's more than 9,900 degrees Fahrenheit!

13. At its core, the sun is far hotter: 15.7 million kelvin, or 27 million degrees Fahrenheit.

14. Sunspots appear and disappear, and levels of solar radiation rise and fall, in an 11-year cycle.

15. Sunspots are areas on the surface that are a bit cooler than the surrounding areas.

16. They're formed by fluctuations in the magnetic field.

17. When we say a bit cooler, we don't mean humans could live there—sunspots average about 3,800 kelvin instead of 5,800 kelvin.

18. The sun rotates faster at its center than at its poles.

19. The sun releases charged particles—the solar wind—from its upper atmosphere.

20. Auroras are caused by fluctuation and flares in the solar wind, as the particles meet the gases in Earth's atmosphere.

21. In auroras, the gas in our atmosphere determines the color.

22. Red and green are caused by oxygen, while blue and purple are associated with nitrogen.

23. Astronomers on Earth first spotted solar flares in 1859.

24. Richard Carrington identified a geomagnetic storm and tied it to solar activity.

25. Geomagnetic storms caused by solar flares that release electromagnetic radiation can cause power outages on Earth.

26. The sun is almost a perfect sphere.

27. Earth's surface gravity is 9.8 meters per second squared. By comparison, the sun's is 274 meters per second squared.

5 Brightest Natural Objects

1. Sun (brightest star)

2. Moon (brightest natural satellite)

3. Venus (brightest planet)

4. Jupiter

5. Sirius (brightest night star)

30 Mercury Facts

1. Light travels from the sun to Mercury in 3.2 minutes.

2. Mercury travels around the sun every 88 days.

3. If you were on Mercury, the sun would appear three times larger than it does on Earth.

4. Mercury's temperature during the day can rise to 800° F.

5. But it doesn't stay warm at night, when temps drop to -290° F.

6. At the poles, the temperature never rises above the freezing point.

7. It never even gets near it, with a high temp of -135° F.

8. Mercury's orbit around the sun isn't a perfect sphere—it's more like an egg.

9. At its closest point it gets to 29 million miles away from the sun, but at its farthest point it moves to 43 million miles away.

10. You could fit 24,462 planets the size of Mercury, the smallest planet, into Jupiter, the largest one.

11. Mercury can be seen without a telescope, but not always easily.

12. It's easiest to spot near sunrise or dusk.

13. Mercury was first recorded by Assyrian and then Babylonian astronomers.

14. The Greeks were the ones to realize that when they sighted Mercury in the morning, it was the same "star" they spotted in the evening.

15. *Mariner 10* was the first spacecraft to visit Mercury back in the 1970s.

16. *MESSENGER* visited Mercury later, leaving Earth in 2004 and doing a flyby in 2008.

17. *MESSENGER* did a final pass in 2015 before crashing on Mercury.

18. *MESSENGER* was an acronym, standing for MErcury Surface, Space ENvironment, GEochemistry, and Ranging.

19. *MESSENGER* detected water ice deposits at Mercury's north pole.

20. Mercury's core has a high iron content, higher than Earth's. Its core makes up more than half of its mass.

21. Mercury's surface has heavy cratering.

22. There are more than 400 named craters as of 2023.

23. Writers and artists from around the world and many different historical periods are represented.

24. Named craters include Ailey (named after choreographer Alvin Ailey), Angelou (named after author Maya Angelou), and Brooks (named after poet Gwendolyn Brooks).

25. Eons ago, Mercury experienced a meteor strike that blasted a crater 800 miles wide called Caloris Basin.

26. The strike actually raised hills on the other side of the planet.

27. Mercury has wrinkles!

28. Well, it has compression folds from when the interior of the planet began to contract as it cooled and the surface formed "wrinkle ridges" and other folds.

29. Mercury, like Earth, has a magnetosphere.

30. *MESSENGER* detected "magnetic tornadoes."

31 Facts About Galileo

1. Galileo was born in Florence, Italy, in 1564.

2. Galileo pioneered the use of the telescope to observe celestial bodies, advanced our understanding of physics, and famously demonstrated that objects of unequal weight fall at the same speed.

3. In 1542, Nicolas Copernicus had proposed a heliocentric model, in which the planets orbit the sun.

4. Without physical evidence, however, it was widely rejected by authorities and leading scientists.

5. In 1609, Galileo first heard about the invention of a telescope in Holland.

6. Intrigued by this new invention, he set out to build his own.

7. Galileo's designs eventually improved the telescopes' magnification by as much as thirty times, fundamentally changing their use from terrestrial observation to astronomical observations.

8. He observed and drew the mountains and craters of the moon in great detail.

9. Galileo discovered several of the moons of Jupiter. Not everything in the solar system revolved around Earth.

10. He observed the phases of Venus.

11. In 1610, Galileo published a catalogue of his discoveries, the *Sidereus Nuncius*, which made him famous throughout Europe.

12. Armed with his discovery of the phases of Venus, which proved that Venus was orbiting the sun, Galileo began to publicly advocate for a heliocentric system.

13. This was seen as a direct challenge to the Church's authority.

14. In 1615 his writings were submitted to the Roman Inquisition.

15. In spite of the fact that Galileo travelled to Rome to defend the theory, in 1616 the Inquisition declared heliocentrism not only "foolish and absurd in philosophy," but heretical to boot.

16. The Pope ordered Galileo to abandon heliocentrism and stop defending the idea altogether.

17. For ten years, Galileo complied, avoiding the controversy.

18. When a more liberal Pope came to power in 1632, he encouraged Galileo to publish an unbiased work about the arguments for and against heliocentrism.

19. In the resulting work, *Dialogue Concerning the Two Chief World Systems*, Galileo instead openly advocated for heliocentrism.

20. He was summoned to Rome and tried for heresy.

21. Galileo maintained his innocence throughout the six-month trial, until he was threatened with torture if he did not admit his guilt and recant his beliefs.

22. He did recant, but reportedly whispered "and yet it moves" in a final act of defiance.

23. Convicted of being "vehemently suspect of heresy," Galileo was sentenced to house arrest, under which he would remain for the rest of his life.

24. While under house arrest, Galileo wrote one of his finest works, *Discourses and Mathematical Demonstrations Relating to Two New Sciences*.

25. In it, he laid out his findings over the past 30 years of scientific study.

26. The two new sciences were the motion of objects and strength of materials.

27. They laid the foundation for kinematics and material engineering.

28. In the same work, Galileo first proposed the idea that bodies of different mass fall at the same rate (this would be demonstrated by Apollo 15 astronauts on the moon), as well as the behavior of bodies in motion, the principle relativity of motion, and the nature of infinity.

29. Among his numerous contributions to astronomy, Galileo was the first to observe that the Milky Way was composed of stars.

30. The Father of Modern Science died in 1642 at the age of 77.

31. In 2016 the *Juno* spacecraft arrived at Jupiter, carrying a plaque in his honor.

23 Venus Facts

1. Aside from Earth's moon, Venus is the brightest object in Earth's night sky.

2. It is depicted in van Gogh's famous painting *The Starry Night*.

3. The Greeks thought Venus was two objects: a morning star and an evening one, named Phosphoros and Hesperos.

4. Its buttery cloud cover is mostly carbon dioxide mixed with sulfur dioxide.

5. Atmospheric pressure at the surface is equal to being half a mile below Earth's seas.

6. The surface temperature can reach 900° Fahrenheit.

7. In 1962, *Mariner 2* flew by Venus and scanned it.

8. It was the first exploration of another planet.

9. The Soviet Union sent a number of probes beginning in the 1960s.

10. *Venera 3*, which crash-landed in 1966, was the first spacecraft to land on the surface of another planet.

11. Several later Soviet probes did land on the planet and transmit information back.

12. Venus spins in the opposite direction as Earth.

13. A complete rotation—a day—on Venus takes 243 Earth-days.

14. Venus's orbit around the sun is a nearly perfect circle.

15. Venus is highly volcanic, and volcanism shaped its surface.

16. Many of the more than 1,600 volcanoes are now dormant or extinct.

17. Evidence exists that some features may still be volcanically active.

18. Venus has a weaker magnetic field than Earth.

19. Venus might have once had oceans on its surface before the sun's radiation and Venus's atmosphere caused them to evaporate.

20. Venus's clouds are made of sulfuric acid.

21. If you were somehow able to stand on the planet's surface, you wouldn't see a lot of sunlight because of the planet's cloud cover.

22. Wind speeds reach 185 miles per hour there.

23. Venus has no moons now, but it may have had them in the past.

32 Facts About Earth

1. Earth is the only planet not named after a god.

2. The word Earth comes from German and means "the ground."

3. Objects weigh slightly less at the equator than at the poles.

4. The distance between the Earth and the sun is about 93 million miles.

5. Earth's orbit isn't a perfect circle around the sun, though, so the distance can vary from 91 million miles (perihelion) to 94.5 million miles (aphelion).

6. Astronomers were able to calculate the distance from the sun to Earth as far back as the 1600s.

7. Earth is the largest of the "rocky" planets.

8. Our magnetic field can change directions.

9. This happens about once every 400,000 years, but not on a regular schedule.

10. When the magnetic field changes directions, the process lasts a few centuries (during which a compass would be completely unreliable) before things stabilize.

11. Earth is the densest planet in our solar system.

12. While Earth only has one moon, it has a lot of other objects orbiting it.

13. There are at least 22,000 "near-Earth asteroids."

14. Of those, about 2,000 pose danger to Earth because of their size and relative closeness.

15. One of those near-Earth asteroids, 3122 Florence (named after Florence Nightingale), even has two small moons of its own!

16. In 2017, Florence passed Earth from a "close" distance of 4,391,000 miles.

17. NASA estimates that about once every 100 years, a meteorite substantial enough in size to cause tidal waves wallops Earth's surface.

18. About once every few hundred thousand years, an object strikes that is large enough to cause a global catastrophe. So future hits—both larger and smaller—are inevitable.

19. The Ensisheim Meteorite, the oldest recorded meteorite, struck Earth on November 7, 1492, in the small town of Ensisheim, France.

20. The Tunguska Meteorite, which exploded near Russia's Tunguska River in 1908, is still the subject of debate more than one hundred years later.

21. It didn't leave an impact crater, which has led to speculation about its true nature.

22. The Hoba Meteorite, found on a farm in Namibia in 1920, is the heaviest meteorite ever found.

23. Weighing in at about 66 tons, the rock is thought to have landed more than 80,000 years ago.

24. Over the years, erosion, vandalism, and scientific sampling have shrunk the rock to about 60 tons.

25. Santa had to compete for airspace on Christmas Eve 1965, when Britain's largest meteorite sent thousands of fragments showering down on Barwell, Leicestershire.

26. Arizona would be short one giant hole in the ground if it wasn't for a 160-foot meteorite landing in the northern desert about 50,000 years ago.

27. It left an impact crater about a mile wide and 570 feet deep.

28. At 186 miles wide, Vredefort Dome in South Africa is the site of the biggest impact crater on Earth.

29. The Sudbury Basin in Sudbury, Ontario, is a 40-mile-long, 16-mile-wide, 9-mile-deep crater caused by a giant meteorite that struck Earth about 1.85 billion years ago.

30. It's also one of the most profitable—the metals from the 6- to 12-mile-wide asteroid that caused the crater brought a motherlode of nickel, copper, and platinum, making Sudbury a metal haven.

31. The 105-mile-wide Chicxulub crater in Yucatán Peninsula, Mexico—an impact crater identified by geologists in 1992—is thought to be associated, or at least partially associated, with the extinction of the dinosaurs.

32. More than 150 impact craters have been identified on Earth, some on the surface and some hidden below the surface.

30 Constellation Facts

1. The word constellation comes from the Latin *cōnstellātiō*, meaning "set with stars."

2. Smaller groupings of stars are known as asterisms.

3. One example of an asterism is Venus' Mirror in the constellation Orion.

4. According to the International Astronomical Union (IAU), there are 88 officially recognized constellations.

5. Forty-eight of these constellations are known as ancient or original, meaning they were created and used by ancient civilizations.

6. The earliest evidence of constellations is found on Mesopotamian clay writing tablets dating back to 3000 B.C.

7. This is likely the origin of Ancient Greek constellations.

8. Ancient constellations were recorded in Greek poet Aratus's *Phenomena* and Greek astronomer Ptolemy's *Almagest*.

9. Modern constellations, like the Peacock, Phoenix, and Sextant, were identified by astronomers of the 14th, 15th, and 16th centuries.

10. Your location on Earth determines what constellations you'll be able to see in the night sky, as well as how high they appear above the horizon.

11. Circumpolar constellations are constellations that never set below the horizon as viewed from a particular latitude.

12. Instead, they make a full circle around a celestial pole, like the North Star in the Northern Hemisphere.

13. The time of year also determines what constellations are visible.

14. Stars appear to move west across the sky during the year as the Earth revolves around the sun.

15. The brightest stars in a constellation are not necessarily the largest.

16. A star's brightness is also dependent on its temperature and distance from Earth.

17. Because each star in a constellation moves independently of one another, all constellations slowly change shape over time.

18. Constellations assist navigators in locating certain stars.

19. This is called celestial navigation.

20. It is especially useful when traveling across the ocean where other landmarks are not visible.

21. NASA astronauts are trained in celestial navigation as a backup in the event their modern instruments fail.

22. Astronomers drew a boundary around each of the 88 official constellations, dividing the celestial sphere into 88 pieces.

23. Any star inside a constellation boundary is considered part of the constellation even if it is not actually part of the image.

24. Constellations are used by astronomers when naming stars and meteor showers.

25. This is helpful when conveying approximate locations because any given point in the sky lies in one of the modern constellations.

26. Astrology is the belief that the location of celestial objects like stars in constellations can describe a person's character or predict the future.

27. There are 12 astrological constellations of the zodiac, but 13 astronomical zodiac constellations: Capricornus,

Aquarius, Pisces, Aries, Taurus, Gemini, Cancer, Leo, Virgo, Libra, Scorpius, Sagittarius, and Ophiuchus.

28. The cycle of the zodiac was used by ancient cultures to determine the time of year.

29. Due to a phenomenon called precession, Earth wobbles on its axis as it rotates.

30. This affects seasonal timing, causing them to shift with respect to zodiac constellations.

23 Facts to Know About Tycho

A golden nose, a dwarf, a pet elk, drunken revelry, and astronomy? Read about the wild life of this groundbreaking astronomer.

1. Tycho Brahe was a Dutch nobleman who is best remembered for blazing a trail in astronomy in an era before the invention of the telescope.

2. Through tireless observation and study, Brahe became one of the first astronomers to fully understand the exact motions of the planets.

3. In 1560, Brahe, then a 13-year-old law student, witnessed a partial eclipse of the sun.

4. He was so moved by the event that he bought a set of astronomical tools and a copy of Ptolemy's legendary astronomical treatise, *Almagest,* and began a life-long career studying the stars.

5. Brahe differed from his forebears in his belief that new discoveries in astronomy could be made, not by guesswork and conjecture, but by rigorous and repetitious studies.

6. His work includes many publications and even the discovery of a supernova now known as SN 1572.

7. Brahe became one of the most widely acclaimed astronomers in all of Europe.

8. When King Frederick II of Denmark heard of Brahe's plans to move to the Swiss city of Basle, the king offered him his own island, Hven, located in the Danish Sound.

9. Once there, Brahe built his own observatory known as Uraniborg.

10. Brahe ruled Hven as if it were his own personal kingdom, forcing tenants to supply him with goods and services or be locked up in the island's prison.

11. At one point Brahe imprisoned an entire family—contrary to Danish law.

12. While famous for astronomy, Brahe is more infamous for his colorful lifestyle.

13. At age 20, he lost part of his nose in an alcohol-fueled duel (reportedly using rapiers while in the dark) after a Christmas party.

14. Portraits of Brahe show him wearing a replacement nose possibly made of gold and silver and held in place by an adhesive.

15. When his body was exhumed in 1901, green rings discovered around Brahe's nasal cavity led some scholars to speculate that the nose may actually have been made of copper.

16. While a considerable amount of groundbreaking astronomical research was done on Hven, Brahe also spent his time hosting legendarily drunken parties.

17. Such parties often featured a little person named Jepp who dwelled under Brahe's dining table and functioned as something of a court jester.

18. Brahe may have believed that Jepp was clairvoyant.

19. Brahe kept a tame pet elk, which stumbled to its death after falling down a flight of stairs—the animal had gotten drunk on beer at the home of a nobleman.

20. He was ostracized for marrying a woman from the lower classes, considered shameful for a nobleman such as Brahe.

21. Due to his shameful union, all of his eight children were considered illegitimate.

22. According to legend, Brahe died from a bladder complication caused by not urinating, out of politeness, at a friend's dinner party where prodigious amounts of wine were consumed.

23. The tale lives on, but recent research suggests this version of Brahe's demise could be apocryphal: He may have died of mercury poisoning from his own fake nose.

15 Exoplanet Facts

1. An exoplanet is any planet outside our solar system.

2. There are billions of them, maybe as many as there are stars.

3. As of January 2023, there are more than 5,295 confirmed exoplanets.

4. Many of those planets have been discovered by TESS, the Transiting Exoplanet Survey Satellite.

5. Most planets, as in our solar system, orbit a star.

6. Those that do not are known as rogue, nomad, or sunless planets.

7. The first exoplanets were discovered in the 1990s.

8. 51 Pegasi B, a gas giant 50 light-years from Earth, was discovered in 1995.

9. It was the first confirmed exoplanet.

10. Only a few exoplanets can be seen by telescopes.

11. Most are determined when people spot stars getting dimmer because the planet passes in front of it, or when a star's light changes in certain ways.

12. Exoplanets are divided into gas giants (like Jupiter); Neptune-like; Super-Earth (rocky planets larger than Earth); and Terrestrial (rocky planets Earth's size or smaller).

13. Some gas giant exoplanets, including 51 Pegasi B, are larger than Jupiter.

14. In 2017, NASA announced the discovery of seven Earth-sized planets orbiting around a star.

15. The planets of this star system, named TRAPPIST-1, possibly contain liquid water and water vapor.

23 Facts About the Moon

1. Our moon is the fifth largest moon in the solar system, after three of Jupiter's and one of Saturn's.

2. The moon orbits the Earth from a distance of about 240,000 miles.

3. The moon's equatorial circumference is about 6,784 miles, compared to the Earth's 24,901 miles.

4. The moon helps stabilize the Earth's rotation on its axis.

5. When a Soviet spacecraft flew by the moon in 1959, it was the first time that humans knew what the far side of the moon looked like.

6. The moon does have an atmosphere—it's just very, very thin.

7. The moon's magnetosphere is very weak, compared to Earth's.

8. Apollo missions have brought back more than 840 pounds of lunar material.

9. NASA's Artemis program plans to send people back to the moon by 2024, including the first woman.

10. The moon has more than 9,100 craters.

11. Asteroids and meteoroids continue to form craters.

12. Some of the craters are more than two billion years old.

13. The largest crater on the moon is the South Pole-Aitken Basin, with measures 1,600 miles in diameter.

14. The Oceanus Procellarum on the moon, a vast lunar plain, might have been formed by an impact crater.

15. If so, it would be even larger, with a diameter of 1,611 miles.

16. The "lunar maria" are plains that were formed by volcanoes.

17. The word "mare" was the Latin word for seas.

18. Lunar eclipses occur when the Earth comes between the sun and the moon.

19. A "supermoon" happens when a full moon or a new moon happens at the same time that the moon and Earth are closest to each other.

20. An Indian mission in 2008 discovered evidence of water on the moon.

21. Further exploration found ice water at the lunar poles, trapped within craters.

22. The moon reaches high temperatures of 260° F, and lower temperatures of -280° F.

23. A point called the Selenean Summit, found on the rim of an impact crater, is the moon's highest point, standing about 35,387 feet above the lunar mean.

10 Biggest Moons

1. Ganymede (Jupiter)

2. Titan (Saturn)

3. Callisto (Jupiter)

4. Io (Jupiter)

5. Moon (Earth)

6. Europa (Jupiter)

7. Triton (Neptune)

8. Titania (Uranus)

9. Rhea (Saturn)

10. Oberon (Uranus)

43 Moon-visiting Men Facts

1. Twenty-four men have visited the moon.

2. Twelve walked on the surface and twelve others reached lunar orbit.

3. The Apollo 8 mission launched in 1968 with astronauts Frank Borman, Jim Lovell, and Bill Anders.

4. Frank Borman was the commander of Apollo 8, a fighter pilot and test pilot.

5. Jim Lovell went to the moon twice, with Apollo 8 and Apollo 13.

6. While he was scheduled to walk on the moon with Apollo 13, the mission's technical difficulties meant he did not.

7. Bill Anders took a famous color image of the earth called "Earthrise."

8. Apollo 10 launched in May 1969, as a test flight for the moon landing.

9. Astronauts John Young, Thomas Stafford, and Eugene Cernan were aboard.

10. John Young and Gene Cernan would later land on the moon as part of the Apollo 16 and Apollo 17 missions, respectively.

11. Stafford later served as Chief of the Astronaut Office and the commander of the Apollo-Soyuz Test Project flight.

12. The crew of Apollo 11 that made the first moon landing in July 1969 consisted of Neil Armstrong, Edwin "Buzz" Aldrin, and Michael Collins.

13. Armstrong and Aldrin walked on the moon, while Collins remained in orbit.

14. When Armstrong made his first spaceflight as pilot of *Gemini 8*, he was the first civilian astronaut to fly in space.

15. Aldrin, a Presbyterian, held a private religious ceremony on the moon.

16. Michael Collins was born in Italy, the son of an Army officer.

17. Charles "Pete" Conrad, Alan Bean, and Richard F. Gordon Jr. manned the Apollo 12 mission in November 1969, with Gordon orbiting the moon while the other men landed.

18. Gordon was slated to go to the surface of the moon himself with Apollo 18, but the mission was scrubbed due to budget cuts.

19. Conrad later commanded the first crewed Skylab mission in 1973.

20. Alan Bean painted as a hobby, and later painted pictures of the moon, using real dust gathered from his keepsake patches.

21. Jim Lovell, Jack Swigert, and Fred Haise were the Apollo 13 crew.

22. They spent four days in lunar orbit in a very cold lunar module, *Aquarius*.

23. Fred Haise was slated to return to the moon with Apollo 19, but the mission was cancelled for budgetary reasons.

24. Jack Swigert later ran for office and won his congressional election, but died of cancer before serving.

25. The crew of Apollo 14, who traveled to the moon in 1971, consisted of Stuart Roosa, Alan Shepard Jr., and Edgar Mitchell.

26. Roosa stayed in orbit while the other men went to the surface of the moon.

27. Roosa had been a firefighter with the U.S. Forest Service before joining the Air Force.

28. He carried seeds provided by the U.S. Forest Service with him on the mission, and "Moon Trees" were later grown from those seeds.

29. At age 47, Alan Shepard was the oldest man to walk on the moon, as well as the first to strike golf balls on the surface.

30. Mitchell was later one of the co-founders of the Institute of Noetic Sciences, which studied paranormal phenomena.

31. The 1971 Apollo 15 mission was manned by David Scott, James Irwin, and Alfred Worden; Worden remained in orbit.

32. Irwin experienced irregularities in his heart rhythm during his time on the moon, although it had settled into a regular rhythm by his return to Earth.

33. He later had several heart attacks, including one just a few years later.

34. While Worden was in the command module *Endeavour*, he was 2,235 miles away from his crewmates, earning the distinction of most isolated human being in history.

35. Scott had been the command module pilot of Apollo 9 mission in March 1969.

36. In 1972, John Young, Charles Duke, and Ken Mattingly comprised the crew of the Apollo 16 mission, with Mattingly staying in orbit while the other men landed.

37. At 36 years old, Charles Duke was the youngest man to walk on the moon.

38. Mattingly had been scheduled for the Apollo 13 mission, but was removed from that mission due to exposure to rubella.

39. Young had served as a naval aviator in the Korean War.

40. The final mission to the moon, Apollo 17, took place in December 1972, and was crewed by Eugene "Gene" Cernan, Harrison Schmitt, and Ron Evans.

41. As he re-entered the Apollo Lunar Module after Harrison Schmitt on their third and final lunar excursion, Cernan became the last person to step foot on the moon.

42. Evans, who stayed in orbit while the other men landed, performed an EVA during the return to Earth.

43. It was the final spacewalk of the Apollo program.

25 Facts About Jupiter

1. It takes 43.2 minutes for the light from the sun to reach Jupiter, across an average distance of 484 million miles.

2. It takes about 12 Earth-years for Jupiter to orbit the sun.

3. The surface gravity of Jupiter is more than two-and-a-half times greater than that of Earth.

4. Ammonia and water in the planet's atmosphere appear as swirls and markings through a telescope.

5. The main components of the planet's atmosphere are hydrogen and helium.

6. As of 2023, Jupiter has 92 known moons, though only 57 are named.

7. All of Jupiter's moons, together with smaller moonlets, form a satellite system called the Jovian system.

8. Jupiter's four largest moons (Io, Europa, Ganymede, and Callisto) are some of the largest in the solar system.

9. Ganymede is the biggest. It's even larger than Mercury!

10. In fact, Ganymede is the ninth largest object in our solar system.

11. It is the only moon we've discovered with a magnetic field.

13. The moon Io is considered to be the most geologically active object in our solar system.

14. Astronomers have watched some of Io's 400-plus active volcanoes erupt, and a few of its mountains rise higher than Mount Everest.

15. On March 2, 1998, the spacecraft *Galileo* discovered a liquid ocean on Jupiter's moon Europa.

16. The surface of the moon Callisto is shaped by impact craters.

17. Jupiter has an ocean—but not of water. It is made of liquid hydrogen.

18. Jupiter does have rings, but they're made of dust and pale in comparison to Saturn's.

19. They were spotted by *Voyager 1*.

20. Jupiter's diameter is more than 88,840 miles across (the diameter of Earth is around 7,918 miles).

21. *Pioneer 10* was the first spacecraft to perform a flyby.

22. Jupiter rotates in about 10 hours, giving it the shortest day of any planet.

23. Winds on Jupiter can reach speeds of more than 330 miles per hour.

24. The Great Red Spot, a storm, has been observed for at least three centuries.

25. Jupiter has a powerful magnetic field, significantly more powerful than Earth's, that causes beautiful aurora.

28 Martian Facts

1. It takes 12.6 minutes for light from the sun to reach Mars.

2. Mars's red appearance comes from the iron oxide on its surface.

3. Mars, like Earth, has seasons.

4. Mars orbits the sun in 687 Earth-days.

5. Its average distance from the sun is 142 million miles.

6. At its further point, Mars can be 154 million miles from the sun.

7. At its closest point, Mars can be 128 million miles from the sun.

8. Olympus Mons, the largest volcano in the solar system, rises 16 miles above the Mars's surface—three times higher than Mount Everest—and is the size of Arizona.

9. The caldera (central crater) alone is large enough to swallow a medium-sized city.

10. It's a shield volcano (like the ones in Hawaii), the result of lava flowing out over the ages.

11. The Valles Marineris canyon stretches more than 3,000 miles, with its greatest depth sinking to more than 4 miles.

12. Mars is, of course, named after the Roman god of war.

13. Mars's two moons, Phobos and Deimos, are named after the twin sons of the Greek god of war, Ares, and Aphrodite.

14. The two named craters on Deimos are named Swift and Voltaire.

15. Voltaire wrote a short story in which he suggested that Mars had two moons.

16. Jonathan Swift suggested the same in *Gulliver's Travels*.

17. Areology is the study of the geology of Mars, from the Greek god of war, Ares.

18. The Mandarin name for Mars translates to "fire star."

19. *Mariner 4*, launched in 1964, was the first spacecraft to get close to Mars.

20. The first unsuccessful landing was by the Soviet *Mars 3* mission in 1971.

21. The U.S.'s *Viking 1* mission achieved a successful landing.

22. February of 2021 was a busy month for spacecraft arriving at Mars: the United Arab Emirates sent an orbiter called *Hope*; China sent an orbiter, a lander, and a rover; and NASA sent the *Perseverance* rover.

23. *Perseverance* began to collect the first samples of the Martian surface.

24. Mars is a colder planet than Earth, with a low temperature of -225° F.

25. Even at its warmest, Mars doesn't often exceed 70° F.

26. Mars probably used to have a magnetic field, but doesn't anymore.

27. The planet has ice caps at its poles.

28. Mars has large dust storms that affect the entire planet.

16 Facts About Stars

1. The brightest star in our sky (after the sun, of course) is Sirius.

2. In Greek, the word means "glowing."

3. Sirius is actually a binary star—a star system of two stars, in this case Sirius A and Sirius B.

4. Sirius will only become brighter over time, as it's moving closer to our solar system.

5. Sirius is found in the constellation Canis Major.

6. After Sirius, Canopus is the brightest star in the night sky.

7. Canopus is part of the constellation Argo.

8. Polaris, the Pole Star, is about 433 light-years away.

9. Polaris is a triple star, made up of a yellow supergiant and two smaller stars.

10. Red dwarf stars—the smallest stars—are the most numerous stars in the galaxy.

11. "Hypergiants" are massive stars that are larger, hotter, brighter, and emit more energy than our sun.

12. They are rare and relatively short-lived, with a life span of "only" a few million years.

13. Betelgeuse, seen in the constellation Orion, is a red supergiant.

14. The oldest stars are believed to be about 13.7 billion years old.

15. The color of a star reflects its temperature.

16. We often think of blue as a "cool" color and red as "hot," but actually blue stars are the hottest and red the coolest.

22 Saturn Facts

1. You can easily view Saturn's rings with a basic telescope.

2. Every so often, the planet of Saturn appears to lose its rings.

3. That has to do with their position at the time, when we're only seeing the edge directly. (Think of how thin a record might look if you were looking only at its edge.)

4. Seven hundred planets the size of Earth would fit in Saturn, while about 1,300 would fit in Jupiter.

5. Saturn is about 886 million miles from the sun.

6. It takes about 29 Earth-years for Saturn to orbit the sun.

7. Like Jupiter's atmosphere, Saturn's is made primarily of hydrogen and helium.

8. Saturn has seven rings made of ice and rock.

9. Some of the particles in Saturn's rings aren't very small—think the size of a building rather than a boulder.

10. Saturn is not a very dense planet. It's less dense than water!

11. It takes sunlight 79 to 80 minutes to travel from the sun to Saturn.

12. Some of Saturn's winds can reach speeds of 1,600 feet per second.

13. The planet has a very strong magnetic field.

14. Saturn has a big moon collection: There are 63 named moons as of January 2023, but it likely has 83.

15. Its largest moon, Titan, has an atmosphere (mostly nitrogen with some methane) and standing liquid on its surface.

16. Titan is much bigger than Earth's moon and is the second largest known moon in the solar system, after Jupiter's Ganymede.

17. Saturn's moon Enceladus is covered with ice.

18. Astronomer William Herschel discovered Enceladus.

19. Enceladus is home to cryovolcanoes—volcanoes that shoot water and other materials instead of lava.

20. *Pioneer 11* first visited Saturn in a flyby in September 1979.

21. *Cassini* studied the planet from orbit for 13 years and carried the Huygens probe that landed on Titan.

22. *Cassini* ended its mission by plunging through Saturn's rings and falling into Saturn's atmosphere, achieving purposeful vaporization.

26 Facts About Uranus

1. Uranus is the farthest planet from Earth visible to the naked eye, but you have to know exactly where to look.

2. William Herschel discovered the planet in 1781 using a telescope.

3. At first, he thought it might be a comet.

4. Herschel called his discovery Georgium Sidus, or George's Star, after his monarch King George III.

5. It was later named Uranus to fit in with the other planets.

6. The name was suggested by Johann Bode, another astronomer who had confirmed Herschel's observations.

7. The element uranium was discovered in 1789, and was named after the planet.

8. It takes Uranus 84 Earth-years to orbit the sun.

9. Uranus has a small rocky core surrounded by water, methane, and ammonia.

10. Its atmosphere consists primarily of hydrogen and helium.

11. Small amounts of methane in the atmosphere give the planet a blue-green color.

12. Despite it being an "icy" planet, scientists estimate that the temperature at Uranus's core is around 9,000° F.

13. The odd thing about Uranus is that it's oriented on its side.

14. Earth is tilted 23.5 degrees out of the plane of the solar system (our local "up" reference point in space).

15. Uranus is tilted 98 degrees, so its poles are in the middle and its equator runs from top to bottom.

16. Uranus has at least 13 rings.

17. It may have more that are yet to be discovered.

18. The only spacecraft to visit Uranus has been *Voyager 2*, and it was only a flyby that occurred in 1986.

19. Uranus has 27 known moons.

20. Many are named after Shakespearean characters, including Miranda and Ariel.

21. The largest moon is Titania, which may contain water.

22. It takes sunlight nearly 160 minutes to reach Uranus.

23. Wind speeds on Uranus can reach 560 miles per hour.

24. While most of Uranus's rings are gray, one is reddish and another is blue.

25. The rings are likely younger than the planet itself, forming later.

26. Uranus has a magnetosphere.

27 Edwin Hubble Facts

We've all heard of the Hubble Space Telescope, the high-resolution telescope that has sent us so many amazing visuals. But who, exactly, was Hubble?

1. Edwin Hubble was born in Marshfield, Missouri, in 1889.

2. He didn't set out to be a scientist, even though astronomy was a hobby when he was a boy.

3. Instead, he studied law at the University of Chicago and The Queen's College, Oxford, fulfilling a promise to his father, who preferred he study law.

4. But after his father died in 1913, Hubble went back to school, studying astronomy at the University of Chicago's Yerkes Observatory, where he earned a PhD in 1917.

5. In a foretelling of things to come, his dissertation was entitled, "Photographic Investigations of Faint Nebulae," for which he used of one of the world's most powerful telescopes at the time, housed at the observatory.

6. After serving in the U.S. Army during World War I, Hubble began working at the Carnegie Institution for Science's Mount Wilson Observatory, where he worked for the rest of his life.

7. Hubble's colleagues espoused theories about the cloudy patches of nebulae they observed through the institution's 100-inch Hooker Telescope, then the world's largest.

8. The prevailing theory at the time was that the Milky Way galaxy comprised the entire universe.

9. So these nebulae, Hubble's fellow astronomers believed, were all located within the galaxy.

10. But when Hubble began studying the nebulae his colleagues had been observing, he wasn't so sure that they were all confined to the Milky Way.

11. One of these nebulae, named the Andromeda nebula, contained a very bright star called a "Cepheid variable."

12. These stars pulsate in a predictable way that can help to determine their distance from an observer.

13. Over several months in 1923, Hubble took pictures of the Andromeda nebula and measured the brightness of the Cepheid variable star, determining that its brightness varied over a period of 31.45 days.

14. He also calculated that it was 7,000 times brighter than our own sun, and determined that it was 900,000 light-years away.

15. Scientist Harlow Shapley, who headed the Harvard College Observatory from 1921 to 1952, had just a few years prior calculated the distance across the Milky Way to be about 100,000 light-years.

16. Although Shapley at first criticized Hubble's findings, these measurements clearly showed that the Andromeda nebula was located far outside of the Milky Way galaxy, proving that the universe was much vaster than previously assumed.

17. Today, scientists know that there are actually two types of Cepheid variable stars, and Hubble's calculations were a bit off: the Andromeda galaxy, as it's now called, is approximately two million light-years away.

18. As telescopes advanced, Hubble began to observe more galaxies, measuring their brightness and noticing a shift in the light toward the red end of the spectrum.

19. This "redshift" occurs due to the Doppler effect, which is a change in the frequency or wavelength of a wave relative to movement.

20. Hubble measured as many galaxies as he could, and in 1929 published a paper revealing his findings.

21. The universe, Hubble realized, was not a static cosmos as had been assumed, but rather it was moving outward.

22. All of the galaxies we can observe are flying apart from each other at great speeds, causing our universe to expand.

23. Hubble calculated that each galaxy is moving at a speed in direct proportion to its distance, a concept now known as Hubble's Law.

24. Galaxies that are farther away are moving more quickly than galaxies that are closer to us.

25. By measuring the rate of expansion, scientists determined the approximate age of the universe, which is about 13.8 billion years old.

26. Scientists also realized that if the universe is continually expanding, that must also mean that at one time, it was much smaller and more compact.

27. This realization eventually gave rise to the Big Bang theory.

21 Facts About Neptune

1. Johanne Galle discovered Neptune in 1846, using sophisticated mathematical predictions.

2. The planet's existence had been deduced earlier than that by other astronomers, based on Uranus's orbit.

3. One of those astronomers, Frenchman Urbain Le Verrier, later tried to name the planet after himself. He did not succeed.

4. You can't see Neptune without a good telescope.

5. It's dimmer than Jupiter's larger moons.

6. Neptune takes 165 Earth-years to get around the sun once.

7. It takes more than four hours for light from the sun to reach Neptune.

8. The light that reaches it is very dim.

9. Neptune appears blue because its atmosphere contains methane gas.

10. The blue of Uranus and the blue of Neptune are different shades, so Neptune may have something else in its atmosphere interacting with the methane.

11. Methane, along with water and ammonia, form the planet's substance.

12. Scientists dub Neptune an "ice" planet for that reason, though the fluid mixture is actually hot and dense.

13. The planet likely has a core of iron and nickel.

14. Neptune is the windiest planet in the solar system.

15. Winds on Neptune can reach 1,200 miles per hour.

16. At least one storm on its surface, titled the "Great Dark Spot," was large enough to contain Earth.

17. It has since dissipated.

18. Neptune has 14 known, named moons as of January 2023.

19. Its main moon, Triton (a little smaller than our own), is doomed.

20. In as little as 10 million years, when Triton falls from the sky and spirals toward Neptune, the planet's gravity will crumble it into a huge ring.

21. Like Uranus, Neptune has only been visited by a *Voyager 2* flyby in 1989.